TEXTBOOK OF ANIMAL BIOTECHNOLOGY

Dr. P.R. Yadav
Lecturer
Dept. of Zoology
D.A.V. College
Muzaffarnagar (U.P.)
(India)

D P H

DISCOVERY PUBLISHING HOUSE PVT. LTD.
NEW DELHI-110 002

[Responsibility for the facts stated, opinions expressed, conclusions reached and plagiarism, if any, in this volume is entirely that of the Author/Editor. The Publishers bear no responsibility for them, whatsoever.]

First Published-2009

ISBN 978-81-8356-495-3

© Author

Published by:

DISCOVERY PUBLISHING HOUSE PVT. LTD.
4831/24, Ansari Road, Prahlad Street
Darya Ganj, New Delhi-110002 (India)
Phone: 23279245 • Fax: 91-11-23253475
E-mail: dphbooks@rediffmail.com
dphtemp@indiatimes.com
web: www.discoverypublishinggroup.com

Printed at:
Sachin Printers
Delhi

Preface

Biotechnology is rapidly acquiring a prominent place in animal research. Animal biotechnology worldwide has made rapid advances in production and health, in the livestock as well as fish industries. The application of new biotechnology methods in livestock industry, such as recombinant DNA techniques, cell culture, monoclonal antibody and others, have already generated a number of products for improving milk and meat production, animal health and food processing, and will definitely continue to do so. Products now emerging ranged from rapid diagnostic tests for contaminants in animal products to genetically engineered animals to produce high value pharmaceuticals. In the veterinary medical research, biotechnology was initially applied to vaccine development and diagnostics for early disease detection. Recently this novel technology has contributed toward efforts for improvement of detoxification systems.

Animal biotechnology is the application of scientific and engineering principles to the processing or production of materials by animals or aquatic species to provide goods and services (NRC 2003). Examples of animal biotechnology include generation of transgenic animals or transgenic fish (animals or fish with one or more genes introduced by human intervention), using gene knockout technology to generate animals in which a specific gene has been inactivated, production of nearly identical animals by somatic cell nuclear transfer (also referred to as clones), or production of infertile aquatic species.

Biotechnology is a rapidly growing and dynamic industry spanning many disciplines in science and engineering. It has been acknowledged by many that biotechnology is the engine of growth for life sciences in the 21st century. India has wisely invested in biotechnology to achieve a rapid advancement in agriculture, human health and other relevant industrial sectors. Since its establishment in May 1995, the National Biotechnology Directorate (BIOTEK), Ministry of Science, Technology and the Environment has played a major role in spearheading national biotechnology programme via Research, Development and Commercialisation, Corporate and Communication, and Human Resource Development.

—**Author**

Contents

1. **Introduction**   1
   Application of animal biotechnology; Transgenics; Gene Knockout Technology; Somatic Cell Nuclear Transfer

2. **Gene Knockout Technology**   8
   Parameters Affecting the Efficiency of Targeted Mutagenesis in Bovine Cells; Objectives; Targeted Mutagenesis; Fish Embryo Cell Cultures for Targeted Gene Inactivation

3. **Cloning with Somatic Cell Nuclear Transfer**   15
   Introduction; Materials and Methods; Nt Embryos; Ivp Embryos; Artificial Insemination; Pregnancy Rates; Maternal Breeds and Pregnancy Losses; Morphometric Analysis

4. **In Vitro Fertilisation and Cell Culture**   39
   Introduction; The Ivf Cycle; Cell Culture; The Growth of Cell Culture; Organ Culture; Techniques of Organ Culture

5. **Breeding of Transgenic Animals**   46
   Introduction; The Reasons for Transgenic Animal's Production; How are Transgenic Animals Produced?; How do Transgenic Animals Contribute to Human Welfare?

6. **Transgenesis and Gene Therapy**   52
   Introduction; Gene Therapy; Embryonic Stem Cell-mediated Transgenesis; Transgenic Models of Astrocytomas; Transgenic Astrocytoma Models; Different Types of Viruses Used As Gene Therapy Vectors; The Recent Developments In Gene Therapy Research

7. **Animal Cloning**   64
   Introduction; Development of Animal Cloning in the Lab; The Process of Animal Cloning; The Types of Cloning Technologies; What Animals have been Cloned?; Can Organs be Cloned for use in Transplants?; What are the Risks of Cloning?

8. **Genetic Engineering (Pharm Animals)**   76
   Introduction; Genetic Modification of Farm Animals; Production of Modified Food-producing Animals; Ethical Issue

9. **Biotechnology for Animal Breeding**   85
   Introduction; Molecular Genetics and Animal Breeding; Molecular Genetics—The Future for Animal Breeding; Genetic Variation; Selection for a Strong Immune Response; New Scientific Technologies; The Future

10. **Priorities in Animal Biotechnology**   94
    Introduction; Animal Models for the Study of Human Health; The Livestock Industries; Development of Rapid Diagnostic Kits; The Priority Areas Identified for Animal Biotechnology

11. **Farm Animal Diversity**   99
    Introduction; Management of Animal Genitic Resources; Some Breeds are more Equal; Hotspot of Breed Diversity Loss; Protecting our Common Heritage; Maintaining Genetic Diversity of Livestock

12. **Farm Animal Genomics**   106
    Introduction; DNA and Protein Sequences Data Banks; Farm Animals: An Unexploited Gold Mine for Biotech; The Swine Sequence; The Bos Taurus Genome; Sheep (Ovis Aries) Genome Project; The Fish Genome Project; The Forgotten Rabbit; Farm Animal Genomics: Current Statistics; Quantitative Trait Loci And Genetic Linkage; Genome Policies for Transgenic Animals; Ethical Reservations of Farm Animal Genomic Study; The Potential of Farm Animal Genomics

13. **Recombinant Dna Technique** 120
   Introduction; Cloning and Relation to Plasmids; Chimeric Plasmids; Synthetic Insulin Production using Recombinant DNA

14. **Anaerobic Digestion in Animals** 127
   Introduction; Odour Impact; Digestor Design; Application; Conclusion

15. **Animal Proteomics** 134
   Introduction; Protein Database of Drosophila; Application of Proteomics to Studies on Phylogeny and Evolution; Antigen Presentation Studies using Mass Spectrometry Techniques; Proteomics on the Totipotent Planarian Stem Cell; Genomics and Proteomic; Cell Proliferation; Applications of Proteomics to Animal Physiology

16. **Animal Cell Biotechnology** 141
   Introduction; Immortalization of Cells in Culture; Exploration on Animal Model for Senile Memory Deficits; Gene Transfer in Animals

17. **Animal Genetic Resources for Agriculture and Food Production** 148
   Introduction; Food Security and Livestock-keys to Poverty Alleviation; Livestock Revolution Underway; Diversified use of Livestock; Diversity in Animal Genetic Resources Invaluable for Future Developments; Considerable Genetic Variation Among Breeds; Why Worry About Loss in Genetic Diversity?; New Approaches Needed for Sustainable Livestock Improvement

18. **Animal Monoclonal Antibody** 162
   Introduction; Monoclonal Antibody Technology; Production of Monoclonal Antibody; Hybridoma Cell Production; Monoclonal Antibodies for Cancer Treatment; Types of Monoclonal Antibodies

19. **Morphological Abnormalities in Animals** 178
   Introduction; Endocrine Disrupting Compounds (EDCs); Cellular and Molecular Mechanism Underlying EDC

Effects; Critical Periods in Development; Experimental Models for Testing EDC Effects; The Biology of the Developing Nervous System in Xenopus Laevis; Conclusion

20. **Applications of Mechatronics to Animal Production**   188

Introduction; Chemotherapeutics in Animal Husbandry; Biosensors Developments; Detection of Estrus in Dairy Cows; Conclusion

Index   *203*

1
Introduction

APPLICATION OF ANIMAL BIOTECHNOLOGY

Animal biotechnology is the application of scientific and engineering principles to the processing or production of materials by animals or aquatic species to provide goods and services (NRC 2003). Examples of animal biotechnology include generation of transgenic animals or transgenic fish (animals or fish with one or more genes introduced by human intervention), using gene knockout technology to generate animals in which a specific gene has been inactivated, production of nearly identical animals by somatic cell nuclear transfer (also referred to as clones), or production of infertile aquatic species.

Animal biotechnology is the application of scientific and engineering principles to the processing or production of materials by animals or aquatic species to provide goods and services (NRC 2003). Examples of animal biotechnology include generation of transgenic animals or transgenic fish (animals or fish with one or more genes introduced by human intervention), using gene knockout technology to generate animals in which a specific gene has been inactivated, production of nearly identical animals by somatic cell nuclear transfer (also referred to as clones), or production of infertile aquatic species.

Since the early 1980s, methods have been developed and refined to generate transgenic animals or transgenic aquatic species. For example, transgenic livestock and transgenic aquatic species have been generated with increased growth rates, enhanced lean muscle mass, enhanced resistance to disease or improved use of dietary phosphorous to lessen the environmental impacts of animal manure. Transgenic poultry, swine, goats, and cattle also have been produced that generate large quantities of human proteins in eggs, milk, blood, or urine, with the goal of using these products as human pharmaceuticals.

Biotechnology is rapidly acquiring a prominent place in animal research. Animal biotechnology worldwide has made rapid advances in production and health, in the livestock as well as fish industries. The application of new biotechnology methods in livestock industry, such as recombinant DNA techniques, cell culture, monoclonal antibody and others, have already generated a number of products for improving milk and meat production, animal health and food processing, and will definitely continue to do so. Products now emerging ranged from rapid diagnostic tests for contaminants in animal products to genetically engineered animals to produce high value pharmaceuticals. In the veterinary medical research, biotechnology was initially applied to vaccine development and diagnostics for early disease detection. Recently this novel technology has contributed toward efforts for improvement of detoxification systems. The important areas of interest by animal researchers worldwide include:

- animal cloning-transgenic animals for the introduction of important traits. The species of animals involved ranged from cattle, sheep, poultry and fish;
- development of rapid diagnostic kits;
- development of recombinant and DNA vaccines;
- improvement of reproductive biotechnology methods;
- treatment of unutilised local feed resources with genetically engineered microbes.

Indian scientists have been researching on various aspects of biotechnology as an alternative means of enhancing the economic production of livestock and fish. Expertise and

Introduction

laboratory facilities in biotechnology and allied disciplines were pooled. Research projects were coordinated under Animal Biotechhology Cooperative Centre, and funded under IRPA. The areas included:

- mass production of in vitro sexed embryos, using new reproductive biotechnologies and recombinant DNA technology;
- development of recombinant vectors for vaccine delivery;
- improvement of genetic resources for better commercial value in fish and wild-life species.

Other basic research in animal biotechnology included isolation and characterisation of local organisms for development of rapid and cost effective diagnostic reagents and kits, vaccines and other biologics.

Animal biotechnology research will involve not only animal researchers from various institutions, but also researchers from other BCCs. The platform technologies such as genomics, transgenic technology, recombinant DNA, biosensor technology and bioprosessing will be utilised. Based on the research progress on the requirements for animal industry, the priority areas identified for animal biotechnology are:

- genetic engineering of animals for improved production and quality;
- development of new generation vaccines and rapid diagnostic kits for animal;
- novel vaccine and drug delivery systems;
- development of cheap feedstuff from local resources;
- improvement of reproductive technologies.

As we progress forward in animal biotechnology research, there are several issues and challenges faced by researchers as outlined under the following SWOT analysis.

Biotechnology is a rapidly growing and dynamic industry spanning many disciplines in science and engineering. It has been acknowledged by many that biotechnology is the engine of growth for life sciences in the 21st century. India has wisely invested in biotechnology to achieve a rapid advancement in agriculture, human health and other relevant industrial sectors. Since its establishment in May 1995, the National Biotechnology Directorate (BIOTEK), Ministry of Science, Technology and the Environment has played a major role in spearheading national biotechnology programme via Research, Development and Commercialization, Corporate and Communication, and Human Resource Development.

TRANSGENICS

Since the early 1980s, methods have been developed and refined to generate transgenic animals or transgenic aquatic species. For example, transgenic livestock and transgenic aquatic species have been generated with increased growth rates, enhanced lean muscle mass, enhanced resistance to disease or improved use of dietary phosphorous to lessen the environmental impacts of animal manure. Transgenic poultry, swine, goats, and cattle also have been produced that generate large quantities of human proteins in eggs, milk, blood, or urine, with the goal of using these products as human pharmaceuticals. Examples of human pharmaceutical proteins include enzymes, clotting factors, albumin, and antibodies. The major factor limiting widespread use of transgenic animals in agricultural production systems is the relatively inefficient rate (success rate less than 10 per cent) of production of transgenic animals. CSREES has supported research projects to generate transgenic animals or transgenic aquatic species with enhanced production or health traits.

GENE KNOCKOUT TECHNOLOGY

Animal biotechnology also can knockout or inactivate a specific gene. Knockout technology creates a possible source of replacement organs for humans. The process of transplanting cells, tissues, or organs from one species to another is referred to as "xenotransplantation." Currently, the pig is the major animal being considered as a xenotransplant donor to humans. Unfortunately, pig cells and human cells are not immunologically compatible. Pig cells express a carbohydrate epitope (alpha 1, 3 galactose) on their surface that is not normally found on human cells. Humans will generate antibodies to this epitope, which will result in acute rejection of the xenograft. Genetic engineering is used to knock out or inactivate the pig gene (alpha 1, 3 galactosyl transferase) that attaches this carbohydrate epitope on pig cells. Other examples of knockout technology in animals include inactivation of the prion-related peptide (PRP) gene that may generate animals resistant to diseases associated with prions (bovine spongiform encephalopathy [BSE], Creutzfeldt-Jakob Disease [CJD], scrapie, etc.). Most of the funding for these types of projects is conducted by private companies or in academic laboratories supported by the National Institutes of Health. Research projects designed to provide basic information regarding mechanisms associated with gene knockout technology are supported by CSREES.

SOMATIC CELL NUCLEAR TRANSFER

Another application of animal biotechnology is the use of somatic cell nuclear transfer to produce multiple copies of animals that are nearly identical copies of other animals (transgenic animals, genetically superior animals, or animals that produce high quantities of milk or have some other desirable trait, etc.). This process has been referred to as cloning. To date, somatic cell nuclear transfer has been used to clone cattle, sheep, pigs, goats, horses, mules, cats, rats, and mice. The technique involves culturing somatic cells from an appropriate tissue (fibroblasts) from the animal to be cloned. Nuclei from the cultured somatic cells are then microinjected into an enucleated oocyte obtained from another individual of the same or a closely related species. Through a process that is not yet understood, the nucleus from the somatic cell is reprogrammed to a pattern of gene expression suitable for directing normal development of the embryo. After further culture and development in vitro, the embryos are transferred to a recipient female and ultimately will result in the birth of live offspring. The success rate for propagating animals by nuclear transfer is often less than 10 per cent and depends on many factors, including the species, source of the recipient ova, cell type of the donor nuclei, treatment of donor cells prior to nuclear transfer, the techniques employed for nuclear transfer, etc. CSREES has supported research projects to obtain a better understanding of the basic cellular mechanisms associated with nuclear reprogramming.

Production of Infertile Aquatic Species. In aquaculture production systems, some species are not indigenous to a given area and can pose an ecological risk to native species should the foreign species escape confinement and enter the natural ecosystem. Generation of large populations of sterile fish or mollusks is one potential solution to this problem. Techniques have been developed to alter the chromosome complement to render individual fish and mollusks infertile. For example, triploid individuals (with three, instead of two, sets of chromosomes) have been generated by using various procedures to interfere with the final step in meiosis (extrusion of the second polar body). Timed application of high or low temperatures, various chemicals, or high hydrostatic pressure to newly fertilised eggs has been effective in producing triploid individuals. At a later time, the first cell division of the zygote can be suppressed to produce a fertile tetraploid individual (four sets of chromosomes). Tetraploids can then be mated with normal diploids to produce large numbers of infertile triploids. Unfortunately, in a commercial production system, it is often difficult to obtain sterilisation of 100 per cent of the individuals; thus, alternative methods are needed to ensure reproductive confinement of transgenic fish. Another technique that is being developed for finfish is to farm monosex fish stocks. Monosex populations can be produced by gender reversal and progeny testing to identify XX males for producing all female stocks or YY males for producing all male stocks. CSREES has supported research projects to alter the chromosome content or produce monosex populations of genetically engineered fish or mollusks.

As with any new technology, animal biotechnology faces a variety of uncertainties, safety issues and potential risks. For example, concerns have been raised regarding: the use of

Introduction

unnecessary genes in constructs used to generate transgenic animals, the use of vectors with the potential to be transferred or to otherwise contribute sequences to other organisms, the potential effects of genetically modified animals on the environment, the effects of the biotechnology on the welfare of the animal, and potential human health and food safety concerns for meat or animal products derived from animal biotechnology. Before animal biotechnology will be used widely by animal agriculture production systems, additional research will be needed to determine if the benefits of animal biotechnology outweigh these potential risks. The USDA Biotechnology Risk Assessment Grants programme supports environmental risk assessment research projects on genetically engineered animals.

2
Gene Knockout Technology

Knockout technology creates a possible source of replacement organs for humans. The process of transplanting cells, tissues, or organs from one species to another is referred to as "xenotransplantation.

PARAMETERS AFFECTING THE EFFICIENCY OF TARGETED MUTAGENESIS IN BOVINE CELLS

Non-technical Summary: Targeted mutagenesis can be used to create genetically modified animals; however, to date, other than one reported success in sheep, there are only published reports of success in mice. The proposed research is aimed at increasing the efficiency of targeted mutagenesis in bovine cells, which could in turn be used to create animals by cloning.

OBJECTIVES

The objectives of this project are to quantify the effects of sequence polymorphisms on the efficiency of gene targeting in bovine cells. This study will make use both of polymorphisms occurring naturally between different alleles of a gene within the species, and of polymorphisms introduced experimentally into an otherwise isogenic targeting vector. To accomplish this, we have defined three specific aims, as follows: 1. Production of a clonal line of bovine embryonic fibroblasts (BEFs) and an isogenic genomic library; 2. Isolation of genomic tbp clones and targeting vector construction; 3. Assessment of targeting efficiency in the BEFs.

Approach: The approach to fulfilling these three specific aims will be the following:

1. Production of a clonal line of bovine embryonic fibroblasts (BEFs) and an isogenic genomic library. Unlike in inbred strains of mice where all genetic loci are homozygous, cattle have a significant degree of heterozygosity. To achieve isogenicity, we will produce a clonal cell line and make the genomic library from this clone. Even though many loci in the clone will be heterozygous, any gene isolated from the library will be isogenic with at least one of the two alleles of that gene in the cell line. After confirming that the cells are diploid, genomic DNA from these cells will be used to produce a l-phage library. To provide a non-isogenic control cell line for comparing targeting efficiency in isogenic and non-isogenic cells, we will produce a second BEF cell line from an embryo from a genetically distant dam and sire.

2. Isolation of genomic TBP clones and targeting vector construction. We have chosen the gene encoding the TATA-binding protein, TBP, on which to test targeting efficiency. The tbp gene is weakly active in all cells, and we have considerable experience mutagenising this gene in mouse cell lines. None of our mutations have had any heterozygous effect on cell growth or viability. Moreover, sequence conservation in the protein-coding region will ease isolation of genomic clones by hybridisation. Genomic clones isolated from our phage library will be restriction-mapped and sequenced. Targeting vectors will contain the neo selectable marker gene and will be designed to facilitate diagnostic screens for gene targeting. To generate "pseudo-non-isogenic" targeting vectors, the targeting vector will be amplified by error-prone PCR and resultant clones will be sequenced. Clones containing between 0.3 and 10 errors per thousand base-pairs will be used tested for gene targeting.

3. Assessment of targeting efficiency in the BEFs. Drug sensitivity in non-transfected cells and drug resistance in cells having a single-copy neo gene will be determined empirically. Once conditions are established, targeting vectors will be transfected into cells by electroporation and transfected cell clones will be selected using these conditions. Individual clones will be screened by Southern blotting and by PCR for correct targeting. Comparison of the efficiency of targeting in the isogenic and non-isogenic cell lines will provide an indication of the importance of sequence identity in targeting BEFs. Comparisons of the targeting efficiency between vectors with different rates of random errors in the isogenic cell line will give a quantitative representation of the effects of polymorphisms on gene targeting in BEFs.

TARGETED MUTAGENESIS

Targeted mutagenesis can be used to create genetically modified animals; however, in species other than mice, there are very few reported successes (one sheep and one pig have been reported in peer reviewed papers) to date. The work performed under this award was aimed at understanding the parameters limiting the efficiency of targeted mutagenesis in bovine cells, which could in turn be used to create animals by cloning. We produced bovine cell lines from a pure-bred breed of dairy cattle (American Holstein) and from an unrelated breed (free-range beef cattle from Montana). DNA from the Holstein cells has been used to produce vectors for targeted mutagenesis. Characterisation of the library indicate it had over 10-fold coverage of the genome with an average insert size of over 22kb. We have screened this library and isolated 5 pure clones of the bovine tbp gene, which will form the target for our studies. This gene has choosen as a representative paradigm largely. Initial studies showed that, since BEFs are highly migratory, they are poor colony formers that were not amenable to clonal selection by traditional drug-selection/colony isolation procedures.

Gene Knockout Technology

The new vector design has both fluorescent and drug-resistance markers. Several days after electroporation, fluorescent cells are isolated by FACS and plated at clonal densities in 96-well dishes. Conditions have been established to give only ~0.66 transfected cells per well, so most colonies are clonal. We have recently begun analysing both polyclonal cell populations and individual clones.

The system is developed for assessing the parameters limiting gene targeting by homologous recombination in bovine cells. Were targeted mutagenesis to become tractable in cattle, it could both increase the value of existing cattle-based commodities and allow creation of countless more valuable cattle-based products. Thus, it could allow production of particularly disease resistant herds or herds exhibiting increased milk or beef production. It could also help create cattle lines custom fit to particular environments, for example, by altering their ability to use different feed sources, altering salt tolerance, or altering reproductive cycles.

Targeted mutagenesis can be used to create genetically modified animals; however, to date, other than one reported success in sheep, there are only published reports of success in mice. The proposed research is aimed at increasing the efficiency of targeted mutagenesis in bovine cells, which could in turn be used to create animals by cloning. Bovine cells are produced lines from a pure-bred breed of dairy cattle (American Holstein) and from an unrelated breed (free-range beef cattle from Montana). DNA from the Holstein cells has been used to produce vectors for targeted mutagenesis. Differences in targeting efficiency between the cell lines will be correlated to the number of DNA sequence differences between the two cell lines at the targeting site. mutation is then introduced into the targeting vector and measure the effects of these on targeting efficiency in the two cell lines. Characterisation of the library indicate the primary library had over 10-fold coverage of the genome with an average insert size of over 22kb. We have screened this library and isolated 5 pure clones of the bovine tbp gene, which will form the target for our studies. Methods were as described for the Holstein line above.

Permanent frozen stocks of these cells have been prepared as above. Preliminary transfection studies have indicated that a modified approach will be required for targeting BEF cells. Thus, since fibroblasts, including BEFs are highly migratory, they are poor colony formers.

Our goal is to measure how critical it is to use isogenic vectors for targeting mutations into bovine cells. Were targeted mutagenesis to become tractable in cattle, it could both increase the value of existing cattle-based commodities and allow creation of countless more valuable cattle-based products. Thus, it could allow production of particularly disease resistant herds or herds exhibiting increased milk or beef production. It could also allow cattle to be used for efficient production of protein-based pharmaceutical products, such as human insulin, human growth hormone, human blood clotting factors, and others.

FISH EMBRYO CELL CULTURES FOR TARGETED GENE INACTIVATION

Non-technical Summary: To optimize aquaculture production, a basic understanding of molecular mechanisms controlling fish development, growth and survival are needed. Cell lines from this project will be used to develop gene transfer technology needed to study the function of specific gene products controlling fish embryo development and growth.

Objectives: Derive a zebrafish primordial germ cell (PGC) line that can be utilized for DNA transfer to the germ line of a host fish embryo. The PGC line will provide a cell-mediated approach to the production of transgenic and gene knockout lines of fish.

Approach: Cell culture methods that were previously developed for in vitro cultivation of fish embryo cells will be optmized to derive a continuously growing PGC line from the zebrafish embryos. The embryo cells will be co-cultured with feeder cells that are able to preserve the embryo cell's ability to produce viable germ cells in vivo.

Continuously-growing ES cell lines were derived from zebrafish embryos at two different developmental stages. The ZEB and ZEG cell lines were derived from blastula- and gastrula-stage embryos respectively, and both lines have been used to generate germ-line chimeras. Germ-line contribution was maintained for a minimum of 6 passages (6 weeks old) and was shown to be stable during freezing and recovery of the cells from liquid nitrogen. Unlike the mouse ES cell culture system, genetic background may not be a limiting factor in generating zebrafish germ-line chimeras. The ZEB and ZEG lines are the first multiple-passage germ-line competent ES cell lines available in any species other than the mouse. In addition to extending the length of time that germ-line competent cells can be maintained in culture, we have also demonstrated that the cells are able to incorporate vector DNA in a targeted fashion by homologous recombination. PCR methods were developed to isolate individual colonies of the homologous recombinants. These results are the first demonstration of homologous recombination in zebrafish cell culture. Work is continuing to use the ES cell cultures for the introduction of targeted mutations into the zebrafish embryo.

The ultimate goal of research is to develop ES cell-mediated gene transfer technology for the production of transgenic and knockout lines of fish. Application of a cell-based gene transfer strategy to agriculturally important fish species will provide a method to efficiently manipulate genetic traits that are important for aquaculture production.

Using the EGFP marker gene, we have developed methods to conveniently identify the germ-line chimeras a few days after introducing the cultured ES cells into host embryos by the presence of fluorescent cells in the developing gonad. Also, we demonstrated that germ-line chimeras can be generated using ES cells and host embryos obtained from zebrafish from at least three different genetic strains. These results indicate that, unlike the mouse ES cell culture system, genetic background may not be a limiting factor in generating zebrafish germ-line chimeras. To further enhance the efficiency of detecting germ-line chimeras generated with the ES cells, we have developed a transgenic line of zebrafish that express the red fluorescent protein (RFP) under the control of the primordial germ cell-specific promoter vasa. This transgenic line will provide a valuable tool that will enable us to accurately identify germ-line chimeras soon after injection. ES cell lines derived from the transgenic embryos will also provide a convenient assay system to optimise the culture system for maintenance of germ-line competency.

The ultimate goal of research is to develop ES cell-mediated gene transfer technology for the production of transgenic and knockout lines of fish. Application of a cell-based gene transfer strategy to agriculturally important fish species will provide a method to efficiently manipulate genetic traits that are important for aquaculture production.

The funded objective of project is to extend the length of time that germ-line competent zebrafish embryo cells can be propagated in culture. During the first year of this project we have derived two continuously growing germ-line competent zebrafish embryonic stem (ES) cell lines. Both cell lines have been used to generate germ-line chimeras. Germ-line contribution was maintained for a minimum of 6 passages (6 weeks old) and was shown to be stable during freezing and recovery of the cells from liquid nitrogen. Germ-line chimeras were generated using ES cells and host embryos from at least 3 different genetic backgrounds. Using ES cells that express the enhanced green fluorescent protein (EGFP) we have developed methods to conveniently identify the germ-line chimeras by the presence of fluorescence in the developing gonad. Also, we demonstrated that the ES cells are able to incorporate vector DNA in a targeted fashion by homologous recombination.

The ES cell lines developed from this work will provide an essential tool for applying gene targeting methods to fish. These techniques will enable researchers to efficiently alter genetic characteristics by targeting inactivation of specific genes in the fish genome. This gene-targeting strategy can be applied to study the function of specific genes that are important for growth and development or to alter characteristics of the aquatic species to improve aquaculture production.

3
Cloning with Somatic Cell Nuclear Transfer

INTRODUCTION

The successful production of viable offspring by cloning with somatic cell nuclear transfer (NT) has been achieved in ungulates, such as cattle, sheep, goat, pig, and mouflon; in rodents, such as the mouse and rabbit ; and in a carnivore, the cat. To date, the technology remains very inefficient and technically demanding. The biggest challenge facing the somatic cell NT technology is the high rate of pregnancy failure as well as fetal morbidity and mortality. Pregnancy failure occurs throughout the entire gestation, from embryo transfer to parturition. Cloned offspring tend to have increased birth weight and poorer post-natal survival and health. However, those that do survive to maturity are fertile and produce normal offspring.

The predominant cause of pregnancy failure associated with NT in cattle and sheep is believed to be either failure of the placenta to form or abnormal placental development and function. Abnormal placental development has also been reported in the cloning of mice. The failure of placental formation may account for most first-trimester losses in cattle, because fetal development is halted in the absence of the placenta. The absence of vascularisation of the chorioallantoic membranes was implicated as a possible cause for the failure of placentation. Abnormal placental development and function probably accounts for a high proportion of post-implantation fetal losses and may contribute to inadequate mammary gland development in the surrogate recipients and the failure of signaling in preparation for parturition. In cattle, the major cause of fetal mortality is the acute, excessive accumulation of allantoic and, to a lesser extent, amniotic fluid, which is collectively referred to as the hydrops syndrome or hydropsy. In the experience of this group, approximately 60 per cent of the fetal losses between Day 120 and full gestation may be attributed to or associated with this syndrome.

The aim of the present study was to compare placental and fetal development in NT cattle pregnancies with closely matched control pregnancies generated either by artificial insemination (AI) or after embryo transfer of in vitro-produced (IVP) embryos. Pregnancies were assessed at the start of placentome formation (Day 50), when the placentomes were completely formed (Day 100), and during the period when hydropsy frequently occurs (Day 150). Contributing factors, such as the genetic background of the fetuses, age of the surrogate recipients and their reproductive history, and nutrition during the pregnancy, were controlled in the experimental design. Use of the same sire allowed comparison of fetal and placental development in the offspring from one sire, although it was not possible to control for maternal gene contribution. The cell line chosen to generate NT embryos had been demonstrated previously to be totipotent, with the production of 20 viable cloned calves. However, previous pregnancies from NT embryos derived from this cell line showed many of the problems associated with clone pregnancies. The present experiment thus allowed us to assess mainly the effect of the NT technology itself on fetal and placental development.

MATERIALS AND METHODS

All manipulations and treatments of animals involved in the present study were conducted in accordance with the regulations of the New Zealand Animal Welfare Act of 1999.

The generation of embryos for transfer to recipients or AI of the heifers was carried out in two separate experiments, conducted approximately 2 mo apart, with equivalent numbers of animals in each experiment. All dams or recipients were 2-year-old heifers of predominantly dairy or dairy cross-breeds. Oocytes were obtained from ovaries of Friesian cows collected at the slaughterhouse, and the same pool was used as a source of recipient cytoplasts for NT or was in vitro fertilised to generate the IVP embryos. In vitro maturation of oocytes was carried out as described previously. To minimise the potential effect of maternal breed on fetal growth and survival, the breed of the recipients was kept as similar as possible for all three groups; however, it was not possible to obtain enough animals from the same breed for these experiments. Only Friesian heifers were inseminated to obtain Friesian fetuses.

NT EMBRYOS

The NT embryos were generated essentially as described previously using the same ovarian follicular cell line (EFC) from a Friesian cow in that study. The cells used for NT in these studies had been passaged at least nine times and previously cryopreserved. Cells were cultured in a 1:1 mixture of Dulbecco modified Eagle medium and Ham F12 (Gibco, Life Technologies, Gaithersburg, MA) supplemented with 10 per cent (v/v) fetal calf serum (FCS; Life Technologies) and 1 mM sodium pyruvate. Cells were induced into a quiescent state by lowering the FCS concentration in the medium to 0.5 per cent and culturing for a further 9 days (experiment 1) or 11 days (experiment 2) before NT. After injection of the donor cells into the perivitelline space, fusion was carried out at 22–24 h after the start of maturation, and fused donor cell/cytoplast couplets were activated 27–28 h after the start of maturation. After activation, in vitro culture was carried out as described by Wells et al. in a biphasic AgResearch Synthetic Oviduct Fluid medium. This medium was replaced on Day 4 of culture with fresh medium containing 10 µM 2,4-dinitrophenol as an uncoupler of oxidative phosphorylation. The fatty acid-free bovine albumin (8 mg/ml) in the AgR SOF was ABIVP. Forty-nine NT embryos judged to be of sufficiently good quality by a subjective grading system were selected for transfer into recipients on Day 7 after fusion. Of the heifers that received NT embryos, 20 of 25 in the first experiment and 11 of 24 in the second experiment were Friesian or Friesian cross-breeds. The remainder were either Jersey cross-breeds or beef cattle.

IVP EMBRYOS

In vitro-matured oocytes were fertilised with frozen-thawed spermatozoa in 50-μl drops under oil for 24 h as described previously. The semen used was from the same Friesian bull that sired the cow from which the EFC cells were isolated. Zygotes were cultured as described for NT embryos, and at Day 7 after fertilisation, the embryos were graded, also as described for the NT embryos. Of those receiving IVP embryos, 8 of 10 from the first experiment and six of nine from the second experiment were Friesian or Friesian-cross. The remainder were beef breeds.

ARTIFICIAL INSEMINATION

Twenty-one Friesian heifers were synchronised for estrus using intravaginal controlled progesterone release devices (CIDR; Pharmacia, Ltd., Auckland, New Zealand) inserted for 12 days. On Day 8, all heifers were injected with 1 ml of estrumate (Schering-Plough, Union, NJ), and the devices were withdrawn after a further 4 days. The mean onset of estrus was approximately 48 h later. Approximately 12 h after the onset of estrus, the heifers were inseminated with frozen semen from the above-described bull used for IVP.

RECIPIENTS OF IVP OR NT EMBRYOS

Seventy healthy heifers were synchronised for estrus concurrent with those heifers used for AI. Single IVP (n = 19) or NT (n = 49) embryos were transferred non-surgically into the uterine lumen ipsilateral to the corpus luteum of each heifer on Day 7 after estrus was observed. The transfer of single embryos avoided the complications of multiple pregnancies.

Pregnancy Monitoring and Morphometric Measurements

After embryo transfer or AI, all heifers were grazed on pasture together. Between Days 40 and 50 of gestation, all heifers were scanned by transrectal ultrasonography using a Piemed 200 scanner with a linear 3.5-to 5-MHz rectal probe (Philipsweg, Maastricht, The Netherlands); pregnant animals were identified and subsequently scanned at monthly intervals until slaughter. Only the presence of a gestational sac was recorded at the first scan. Subsequently, the presence of a fetal heartbeat was regarded as a sign of a viable pregnancy. A sample of pregnant animals from each group was slaughtered at Days 50, 100, and 150 of gestation, and the reproductive tracts were collected and transported back to the laboratory within 1 h. Conceptuses were recovered from the uterus, and morphometric measurements were carried out. The volumes of both allantoic and amniotic fluids were measured, and both fetal weights and crown-rump (C-R) lengths were recorded. The fetal membranes with the cotyledons were weighed wet after all fluids were drained and the membranes separated from the uterus and fetus. At Days 100 and 150, the heart, brain, kidneys, and liver were weighed separately. The appearance of the fetus, organs, uteroplacental units, and fetal membranes were noted. At Days 100 and 150, all caruncles of the uteroplacental units were cut from the uterus after removal of the fetal cotyledons, combined, and weighed, and the numbers were recorded.

Statistical Analyses

At each stage, the treatments were compared for each variable using least significant differences calculated from analyses of variance. Certain variables (fetal brain, heart, liver, and kidney weights) were also expressed as a ratio to the fetal weight. An analysis of variance across stages was done on log-transformed variables to check for treatment by stage interaction. Pairwise comparison of variances between treatment groups was carried out with the F-test, assuming unequal variance. Differences were considered to be significant at $P <= 0.05$.

In Vitro Development of Postfusion NT Embryos

In the first experiment, 81 per cent (105/129) of donor cell/cytoplast couplets were fused successfully, and 104 reconstructed embryos were placed in culture, from which 45 blastocysts (43%) were judged subjectively to be of grades 1 and 2 and, thus, suitable for transfer to recipients. In the second experiment, 83 per cent (108/130) of donor cell/cytoplast couplets were fused successfully. From the 105 of these placed in culture, 45 developed to blastocysts (43%) of grades 1 and 2. Thus, the development rate to grade 1 and 2 blastocysts for the NT embryos was similar in both experiments.

PREGNANCY RATES

The pregnancy rate, as assessed by ultrasonography or at slaughter with the recovery of a fetus, was similar between experiments 1 and 2 for all three treatment groups. Therefore, the data were combined for simplicity of analyses.

Figure 3.1 (*See on next page*) shows the pregnancy rates of the three treatment groups at different stages of gestation. Pregnancy rates after the first slaughter were calculated by taking into account the number that were pregnant at the previous slaughter and adding that to the remaining number that were pregnant at the next ultrasound scanning or slaughter. The data in Fig. 3.1 did not take into account the health and potential viability of the conceptus, only that a conceptus was detected at scanning or slaughter. The pregnancy rates were similar for all three groups at Day 50 (AI, 67%; IVP, 58%; NT, 65%). From then onward, NT pregnancies were continually lost, until only 40 per cent of the recipients that received an NT embryo were still pregnant by Day 150. No fetal losses were recorded with either the AI or the IVP group after Day 50.

MATERNAL BREEDS AND PREGNANCY LOSSES

To determine if maternal breed had any effect on embryonic survival, the conception rates were compared between the groups. At the first ultrasound scan between Days 40 and 50, the conception rate for Friesian heifers undergoing

AI was 66.7 per cent (14/21), which is consistent with the reported mean conception rate of 64 per cent for New Zealand dairy herds after AI. The conception rates for Friesian heifers after embryo transfer was similar for NT (53.6%, 15/28) and IVP embryos (61.5%, 8/13). Conception rate for IVP embryos in cross-bred heifers was 50 per cent (3/6), which is similar to the rate for Friesian heifers. The conception rate for cross-bred heifers after transfer of NT embryos was 81 per cent (17/21), compared with 53.6 per cent for Friesian heifers; however, this difference was not significant (P = 0.09, 2 x 2 exact test).

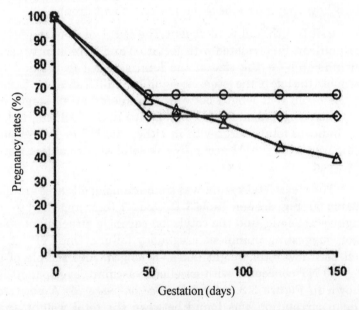

Fig. 3.1: Percentage of animals pregnant after AI () or embryo transfer with single IVP ({diamond}) or NT ({Delta}) embryos. Pregnancy was determined either by ultrasound scanning or with the recovery of a fetus at slaughter

MORPHOMETRIC ANALYSIS

The numbers of AI, IVP, and NT fetuses from which data were obtained at Days 50, 100, and 150 are shown in Table 3.1.

Fetal fluid volumes at Days 100 and 150 sometimes were not recorded, because the tracts had ruptured during slaughter and fluid from the amniotic and allantoic compartments were mixed or, in some cases, partially lost. During estimation of the mean fetal membrane weights, those fetal membranes with large amounts of a gelatinous substance were excluded from the analysis. The mean ± SEM at each stage is presented in Table 3.1 *(See on pages 23 & 24)*. The data are also graphically represented (*See Figs. 3.2 on page 25 and 3.4–3.6 in pages 29, 31, 34*) to display the variability with stage and treatment, with the means for each treatment being joined by a line to show how they relate between the treatment groups.

C-R length and fetal weight The C-R length of the fetus is significantly correlated with gestation and is frequently used an indicator for the size of the fetal skeletal frame and to estimate the age of the fetus. No significant difference was found in the mean C-R length between treatments at all stages of gestation (Fig. 3.2 and Table 3.1). By itself, the C-R length did not indicate fetal overgrowth in either the NT or IVP group compared with the AI group. No skeletal abnormalities were observed.

The mean fetal weight was similar among all three groups at Day 50 (Fig. 3.2 and Table 3.1). One NT fetus appeared to be slightly anaemic, and the cotyledon-caruncle attachment was poor, suggesting imminent pregnancy failure. One IVP fetus had both low fetal weight and C-R length. An example of a Day 50 NT conceptus with excellent vascular development is shown in Figure 3.3, A and B. *(See on page 26)* A positive linear correlation was found between the fetal weight and (C-R length)3 at Day 50 in the AI ($r = 0.79$, $P = 0.11$) and IVP ($r = 0.99$, $P = 0.09$) groups, but the correlation in the NT group was poorer ($r = 0.55$, $P = 0.10$).

Table 3.1: Mean ± SEM valuers for the morphometric measurements made at Days 50, 100 and 150.a

	Gestation days	Treatment AI	Treatment IVP	Treatment NT	Significance of Comparisons[b] NT:AI	NT:IVP	IVP-AI
C-R length (cm)	50	3.97 ± 0.10 (n=5)	3.94 ± 0.45 (n=3)	3.83 ± 0.10 (n=10)	–	–	–
	100	18.78 ± 0.43 (n=4)	18.79 ± 0.27 (n=4)	18.59 ± 0.75 (n=6)	–	–	–
	150	39.99 ± 0.33 (n=5)	38.93 ± 0.73 (n=4)	38.36 ± 0.92 (n=8)	–	–	–
Fetal weight (g)	50	4.43 ± 0.24	4.02 ± 0.75	4.45 ± 0.25	–	–	–
	100	283.4 ± 2.4	302.8 ± 12.2	320.6 ± 31.5	–	–	–
	150	2589 ± 149	2726 ± 196	3042 ± 170	–	–	–
Fetal membrane weight (g)	50	37.2 ± 2.1	46.1 ± 10.1	63.0 ± 6.5	*	–	–
	100	328 ± 36	387 ± 17	304 ± 51 (n=5)	–	–	–
	150	833 ± 117	774 ± 57	1360 ± 271 (n=6)	*	*	–
Total fluid volume (ml)	50	164 ± 13	174 ± 18	213 ± 13	–	–	–
	100	1655 ± 196	1609 ± 204	1206 ± 228 (n=5)	–	–	–
	150	6500 ± 444	5088 ± 698	8033 ± 1800	–	*	–
Allantoic fluid volume (ml)	50	143 ± 11	155 ± 13	194 ± 13	–	–	–
	100	480 ± 213	616 ± 259	235 ± 146 (n=5)	–	–	–
	150	3866 ± 542	1981 ± 751 (n=3)	6433 ± 1531 (n=5)	–	–	–

(Contd...)

	Gestation days	Treatment AI	Treatment IVP	Treatment NT	Significance of Comparisons[b] NT:AI	NT:IVP	IVP:AI
Amniotic fluid volume (ml)	50	20.7 ± 1.8	18.1 ± 4.9	18.9 ± 1.2	–	–	–
	100	1025 ± 38	744 ± 178	935 ± 96 (n=5)	–	–	–
	150	2477 ± 315	2621 ± 265 (n=3)	3985 ± 1021 (n=5)	–	–	–
Total caruncle weight (g)	100	120 ± 11	120 ± 11	214 ± 44	*	*	–
	150	802 ± 28	811 ± 109	1133 ± 86	*	*	–
Caruncle numbers	100	103 ± 15	99 ± 16	58 ± 9	*	*	–
	150	109 ± 10	108 ± 7	98 ± 10	–	–	–
Mean weight per caruncle (g)	100	1.17 ± 0.15	1.21 ± 0.17	3.39 ± 0.67	*	*	–
	150	7.37 ± 0.92	7.48 ± 0.90	11.62 ± 1.52	*	*	–
Brain weight (g)	100	7.43 ± 0.51	8.03 ± 0.80	7.07 ± 0.26	–	–	–
	150	37.2 ± 3.3	39.6 ± 0.7	36.4 ± 1.5	–	–	–
Heart weight (g)	100	2.55 ± 0.08	2.73 ± 0.15	3.33 ± 0.44	–	–	–
	150	22.6 ± 1.53	22.6 ± 2.11	27.1 ± 2.11	–	–	–
Liver weight (g)	100	19.8 ± 0.9	20.2 ± 0.5	17.6 ± 1.2	–	–	–
	150	113 ± 9	105 ± 5	159 ± 10	*	*	–
Kidney weight (g)	100	1.98 ± 0.05	2.10 ± 0.22	2.39 ± 0.42	–	–	–
	150	19.7 ± 0.9	18.8 ± 1.8	27.0 ± 2.8	*	*	–

[a] Sample numbers for the variables at each stage of gestation and treatment are the same as that stated for C-R lengths, unless indicated otherwise.

[b] Means are not significantly different; *, means are significantly different at $P \leq 0.05$.

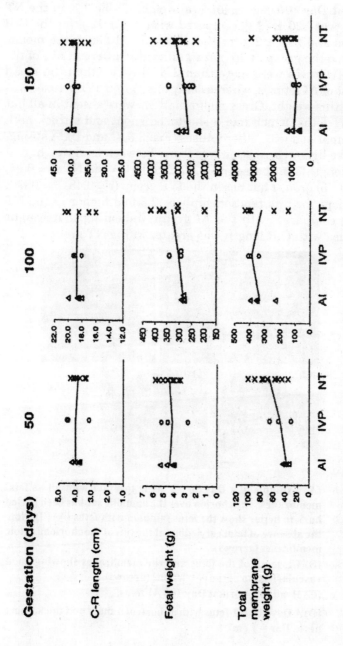

Fig. 3.2: Crown-rump (C-R) length, fetal weight, and weight of combined fetal membranes at Days 50, 100, and 150. The means for each treatment (AI, IVP or NT) are joined by a line

At Day 100, the mean fetal weight (Table 3.1) in the NT group was 320 ± 32 g, compared with 303 ± 12 g for the IVP fetuses and 283 ± 2 g for the AI group. Although the means between the groups were not significantly different, five of the six NT fetuses were more than 2 SD heavier than the mean weight of AI fetuses, whereas only one in four IVP fetuses was above this weight. Gross malformations were absent in all but one NT fetus, which had a shortened snout and a thick neck in comparison with other fetuses (Fig. 3.3, C and D). A strong, positive linear correlation was found between fetal weight and (C-R length)3 in the AI ($r = 0.99$, $P = 0.014$) and NT ($r = 0.94$, $P = 0.005$) groups but not in the IVP group ($r = 0.48$, $P = 0.68$). The slope for this relationship was 7.5-fold higher in the NT group compared with the AI group, indicating that weight increase with C-R length was greater in the NT group.

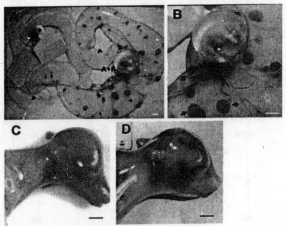

Fig. 3.3. (A) Example of a Day 50 NT fetus together with all its fetal membranes. The chorion over the amniotic sac has been peeled back to better show the fetus (amnion with fetus [A+F]). Note the absence of fetal cotyledons at the tips of the chorioallantoic membranes (arrows).

(B) Close-up of the fetus and the membranes showing good vascularisation of the cotyledons (arrows).

(C) Head of a normal Day 100 AI fetus.

(D) A Day 100 NT fetus with a shortened snout and thick, short neck. Bar = 1 cm

By Day 150, the mean fetal weight had increased 9- to 10-fold in all groups (NT, 3042 ± 170 g; IVP, 2726 ± 196 g; AI, 2589 ± 149 g) Although the mean fetal weight of the NT group was only 17 per cent and 11 per cent higher than those of the AI and IVP groups, respectively (not significantly different), five of the eight fetuses were greater than 2 SD above the mean weight of the AI fetuses, whereas only one in five and one in four were greater than this weight for the AI and IVP groups, respectively. No gross fetal abnormalities were detected in any Day 150 fetus. At this stage of gestation, the skin pigmentation was established, and it was possible to tell that all fetuses were Friesian. No linear correlation between the fetal weight and (C-R length)3 was evident in any of the three groups at Day 150.

Fetal membrane weight— The combined wet weights of the amniotic and chorioallantoic membranes containing the fetal cotyledons are shown in Figure 3.2 and Table 3.1. These fetal membranes are derived entirely from the extraembryonic tissues. At Day 50, all chorioallantoic membranes were vascularised, and the fetal cotyledons were tenuously attached to the uterine caruncles of the uterus. In all cases, the vascularisation and attachment of the cotyledonary burrs to the caruncles were more advanced on the amniochorion and the allantochorion adjacent to the fetus than they were toward the tips of the chorioallantoic sacs. The number of visible cotyledonary burrs was similar in all three groups, with an overall mean of 68 ± 3. The mean fetal membrane weight was significantly greater in NT conceptuses (63.0 ± 6.5 g) compared with the AI group (37.2 ± 2.1 g, $P < 0.05$) but not compared with the IVP group (46.1 ± 10.1 g). The highest membrane weight recorded (101.5 g) was from the anaemic NT fetus, in which attachment of the fetal cotyledons to the caruncles was poor.

The total membrane weight at Day 100 was highly variable, but it was not significantly different between the groups. In two cases in the NT group (one involving the fetus shown in Fig. 3.3D), weights were not recorded because of the extremely gelatinous nature; such gelatinous membranes were absent from either the AI or IVP group. All fetal membranes were well vascularised and the fetal cotyledons firmly attached the caruncles. Two developmentally retarded NT fetuses had correspondingly lower membrane weights. Amniotic pustules, which were located on the inner surface of the amnion facing the fetus, were flat, whitish plaques in the AI group, whereas in several cases in the NT group, they were small, yellow spikes.

By Day 150, the mean fetal membrane weight was significantly higher ($P < 0.05$) in the NT group (1360 ± 271 g) compared with either the AI (893 ± 117 g) or IVP (774 ± 57 g) group. This increased weight may be caused, in part, by a greater total fetal fluid volume in several of the NT pregnancies.

Fetal fluids the total fetal fluid volume represented the combined volumes of the allantoic and amniotic fluids. At all three stages of gestation, no significant difference was found in the mean allantoic, amniotic, or total fluid volume between the treatment groups (*Fig. 3.4 and Table 3.1*). However, 3 of the 10 NT cases at Day 50 and three of the eight cases at Day 150 had total fluid volumes greater than 2 SD above the mean for the AI group at the respective stages of gestation. The ratio of allantoic to amniotic fluid volume varied with gestation. At Day 50, the allantoic fluid volume was 9- to 10-fold higher than the amniotic fluid volume. Increased total fluid volume in all three of the above-described Day 50 NT cases was caused by increased allantoic fluid volume, which was greater than 2 SD above the mean for the AI group.

At Day 100, allantoic fluid volumes were highly variable in all three groups, ranging from 60 to 1300 ml (*See fig. 3.5 on page 31*). The volumes for two NT cases were not recorded, because the tract had ruptured after slaughter in one case and the allantoic sac was full of a jelly-like substance in the other. Amniotic fluid volumes were less variable between individuals. One small NT fetus with only 36 placentomes had low allantoic and amniotic fluid volumes, suggesting a loss of ability to maintain appropriate fetal fluid volume.

Cloning with Somatic Cell Nuclear Transfer

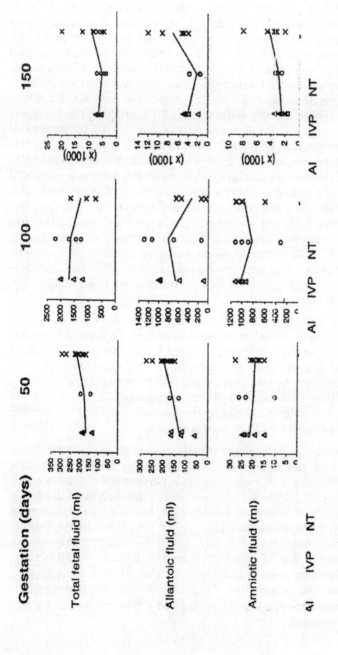

Fig. 3.4: Total fluid, allantoic fluid, and amniotic fluid volumes at Days 50, 100, and 150. The means for each treatment (AI, IVP, or NT) are joined by a line

At Day 150, three NT and one IVP allantoic and amniotic fluid volumes were not recorded, because the fetal membranes had ruptured. Hydropsy was evident in two of eight NT pregnancies and was likely in a third. All three cases had total fluid volumes greater than 2 SD above the AI mean, and two cases also had allantoic fluid volumes that were 2 SD above the AI mean (20 and 12 L compared with 6.6 ± 0.4 L in the AI group). In the animal with 20 L of fluid, the amniotic fluid accounted for 8 L (AI, 2.4 ± 0.3 L), an indication of polyhydramnios, which is far less prevalent than hydrallantois in the hydrops syndrome associated with NT. Increased total fluid volume was associated with increased fetal and placental growth. In two of the above-described cases, fetal weights and heart weights were greater than 2 SD above the AI mean, and the kidney and liver weights in all three cases were 2 SD above the AI mean. All three cases had total caruncle weights greater than 2 SD above the AI mean but normal caruncle numbers. These three cases had the highest kidney and total caruncle weights of any group at this gestation, but the fetuses were not edematous.

Because the allantoic and amniotic fluids are contained within the fetal membranes, we determined if a correlation exists between the fetal membrane weight and the total fetal fluid volume. A positive linear correlation was detected in several instances, but no consistency was observed in this relationship with either treatment or stage of gestation. Thus, it was not always possible to attribute the increase in fetal membrane weight to increased fluid accumulation.

Fetal organ development— The weights of the brain, liver, heart, and kidney were compared among the three groups, both with (Fig. 3.5, expressed as a percentage of fetal weight) and without (Table 3.1) correction for the weight of the fetus. No significant difference was observed in the mean brain weight between groups at either Day 100 or 150. Mean heart weights averaged over Days 100 and 150 were significantly higher ($P < 0.05$) in the NT group compared with either control group; no difference was found between the AI and IVP groups. However, when adjusted for gestation and fetal weight, the difference between the NT group and either control group was not significant.

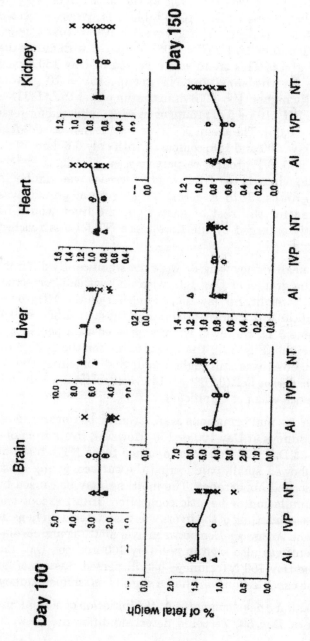

Fig. 3.5: Fetal brain, liver, heart, and kidney weights at Days 100 and 150 expressed as a percentage of the fetal weight. The means for each treatment (AI, IVP, or NT) are joined by a line. At Days 100 and 150, the means for the NT relative liver weight were significantly different from the AI and IVP groups ($P < 0.05$)

No difference was observed in the mean liver weights between the groups at Day 100 (Table 3.1). When corrected for the fetal weight, the livers of NT fetuses were disproportionately lighter ($P < 0.05$) compared with either the AI or IVP group. This trend was reversed by Day 150, when the mean for the surviving NT group (159 ± 10 g) was significantly higher ($P < 0.05$) than that in either the AI (113 ± 9 g) or the IVP (105 ± 5 g) group, even when adjusted for fetal weight ($P < 0.05$). The mean NT liver weight increased 9-fold between Days 100 and 150, compared with only 6.6- and 5-fold in the AI and IVP groups, respectively, suggesting a "catch-up" growth, whereas the fetal weight increase was similar in all three groups (9- to 10-fold). If this rate of growth was maintained for the rest of gestation, the liver would be significantly enlarged by the time these NT fetuses reached full term.

The mean kidney weights were not significantly different between the groups at Day 100, with and without correction for the fetal weight. At Day 150, kidneys from the NT groups were significantly larger ($P < 0.05$) than those in either control group (Table 3.1). However, when expressed as a percentage of fetal weight (Fig. 3.5), the mean value for the NT group ($0.90\% \pm 0.06\%$) was still higher than that for either the AI ($0.77\% \pm 0.10\%$) or IVP ($0.70\% \pm 0.14\%$) control group, although the difference was not significant.

No gross malformations were seen in the brain, liver, heart, or kidneys at Day 100 or 150. However, livers from one NT fetus at Day 100 and from another three NT fetuses at Day 150 showed small, pale foci on the surface, giving these livers a mottled appearance. The mottling may be caused by fatty accumulation or hepatic congestion. In all except one Day 150 case, mottling of the livers was associated with hepatic enlargement. An association between liver mottling and cardiac enlargement was also seen in one Day 100 and one Day 150 case. Three Day 150 NT kidneys had dispersed clusters of fat cells on the capsules, and one had a slightly misshapen kidney.

Placentome development— At the initiation of placentome formation at Day 50, we could detect no difference between the three groups in the number of cotyledonary burrs formed on the chorion. Five of the 10 NT conceptuses had very red fetal cotyledons, suggestive of good vascularisation (*Fig. 3.3*), compared with two from the five AI and none from the IVP group (paler cotyledons).

The number of caruncles and their weights at Days 100 and 150 (*See fig. 3.6 on next page*) were used as indicators of placentome numbers and weights, because the caruncles are the maternal tissues that, together with the fetal cotyledons, form the placentomes. In both the AI and IVP groups, the numbers of caruncles at Days 100 and 150 were very similar (Table 3.1), indicating that placentome numbers were fixed by Day 100. No difference was observed in the caruncle numbers between the AI and IVP groups at either Day 100 or 150. In contrast, the mean number of caruncles at Day 100 in the surviving NT group (58 ± 9) was significantly lower ($P < 0.05$) than in either the AI (103 ± 15) or IVP (99 ± 16) group. Only 36 and 39 caruncles were recorded in two NT pregnancies, and both these fetuses showed signs of growth retardation accompanied by decreased fetal fluid volume and membrane weight. Caruncle numbers in the surviving NT group at Day 150 were similar to those in the two control groups.

Despite lower caruncle numbers at Day 100, the mean weight of all caruncles added together (*Fig. 3.6* and *Table 3.1*) was significantly higher ($P < 0.05$) in the NT pregnancies (NT > AI = IVP). Together with the lower mean caruncle numbers, this resulted in significantly increased ($P < 0.05$) average caruncle weights in this group (NT > IVP = AI). (*See fig. 3.7 on page 36*) shows examples of uterine caruncles from an AI, an IVP, and two NT pregnancies after removal of the fetal membranes and associated cotyledons. The large, flat caruncles in Figure 3.7C were from a uterus in which only 36 were detected, whereas Figure 3.7D shows an NT example with 80 caruncles and increased total and average caruncle weights. A negative linear correlation was found between the caruncle numbers and the average caruncle weight in the AI ($r = 0.91$, $P = 0.09$) and IVP ($r = 0.81$, $P = 0.19$) groups but not in the NT group ($r = 0.26$, $P = 0.62$).

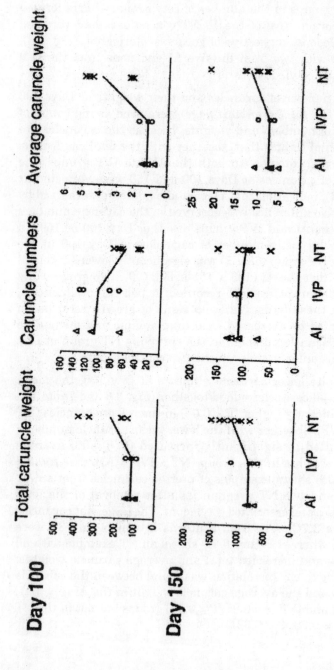

Fig.3.6: Total caruncle weights (g), caruncle numbers, and average caruncle weights (g) at Days 100 and 150. The means for each treatment (AI, IVP, or NT) are joined by a line

Although the NT caruncle numbers were "normal" at Day 150, the mean total caruncle weight was still significantly higher (NT > IVP = AI; $P < 0.05$), resulting in significantly higher ($P < 0.05$) average caruncle weight (NT > IVP = AI). A negative linear correlation between the average caruncle weight and caruncle number was evident in both the AI ($r = 0.98$, $P = 0.002$) and NT ($r = 0.70$, $P = 0.05$) groups but not in the IVP group ($r = 0.11$, $P = 0.88$). The larger of the NT placentomes assumed fist-like structures and were thicker in the sagittal cross-section compared with AI or IVP placentomes (Fig. 3.7E), which were more likely to be flat, discoid structures.

Individual variability—Plots of the morphometric data in Figures 3.2 and 3.4–3.6 demonstrate considerably variability in every parameter examined, particularly within the NT group. We performed F-tests to compare the variances between the groups. Table 3.2 shows the SD for the variables and the significance of the pairwise comparisons of variances between treatment groups. Despite all NT fetuses having the same nuclear genetics of the cell line, a significantly greater variability was observed in the fetal weight at Day 100 than in the contemporary AI ($P < 0.001$) or IVP ($P < 0.05$) group. The IVP group was also more variable compared with the AI group ($P < 0.05$). In addition, the NT kidney weights at Day 100 showed greater variability compared with the AI group ($P < 0.05$); however, this may be caused, in part, by the greater variability in the NT fetal weight. The Day 150 NT liver and kidney weights were more variable compared with those in the AI, but not with those in the IVP, group.

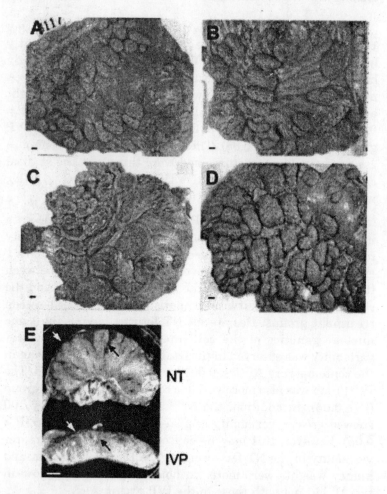

Fig. 3.7: (A–D) Examples of Day 100 uteri from an AI
(A)   with 66 caruncles, an IVP
(B)   with 84 caruncles, an NT with only 36 caruncles
(C)   and an NT with 80 caruncles
(D)   showing the considerably larger NT caruncles. The fetal membranes together with the fetal cotyledons have been removed. E) Sagittal cross-sections of a Day 150 NT and an IVP placentome showing the increased thickening of the NT placentome. The white arrows indicate the fetal side. Fetal villi growing into the caruncle ("streaks") are indicated by the black arrows. Bar = 1 cm

Table 3.2: SD and significance of pairwise comparisons of within-group variances between treatment groups

	SD			Significance of pairwise comparisons[a]		
	AI	IVP	NT	NT:AI	NT:IVP	IVP:AI
Day 100 C-R length (cm)	0.85	0.54	1.83	NS	C	NS
Day 100 fetal weight (g)	409	24.4	77.1	A	C	C
Day 100 kidney weight (g)	0.10	0.44	1.04	B	NS	C
Day 100 heart weight (g)	0.16	0.30	1.08	B	C	NS
Day 150 liver weight (g)	19.6	28.2	50.4	C	NS	NS
Day 150 kidney weight (g)	2.05	8.03	11.5	B	NS	C
Day 50 fetal membrane weight (g)	4.7	17.6	20.6	A	NS	C
Day 100 fetal membrane weight (g)	70.1	33.2	124	NS	C	NS
Day 150 fetal membrane weight (g)	262	114	767	C	B	NS
Day 100 total caruncle weight (g)	21.1	21.0	108	C	C	NS
Day 100 ave. caruncle weight (g)	0.30	0.34	1.34	C	C	NS
Day 150 total caruncle weight (g)	63	217	243	C	NS	C
Day 100 amniotic fluid (ml)	76	357	215	NS	NS	C
Day 150 amniotic fluid (L)	0.70	0.46	2.28	C	C	NS
Day 150 allantoic fluid (L)	1.21	1.30	3.42	C	NS	NS
Day 150 total fluid (L)	0.99	1.40	5.09	B	C	NS

a NS, not significant; A, $p \leq 0.001$; B, $P \leq 0.01$; C $P \leq 0.05$.

Greater variability in the NT group was most evident in the components forming the uteroplacental tissues. Day 50 fetal membrane weights were significantly more variable in the NT group compared with the AI group (P 0.001) but not with the IVP group (P = 0.30). Membrane weights of Day 50 IVP fetuses were also more variable compared with the AI group (P 0.05). At later stages of gestation, some of the variability in fetal membrane weights may be caused, in part, by greater variability in fetal fluid volumes (Table 3.2). By the time placentation was complete at Day 100, the total caruncle and average caruncle weights in the NT group were significantly more variable compared with either the AI (P 0.05) or IVP (P 0.05) group. Total caruncle weights of both the NT (P 0.05) and the IVP (P 0.05) groups were significantly more variable than the weights of the AI group at Day 150. The analyses suggest that greater variability occurs in fetal membrane and placental development in the NT group and, to a certain extent, in the IVP group compared with the AI group.

4

In Vitro Fertilisation and Cell Culture

INTRODUCTION

In vitro fertilisation (IVF) and other "high tech" procedures are now referred to as the assisted reproductive technologies (ART). These procedures all involve collecting the oocytes (eggs) and placing them in direct contact with sperm. Together they form an alphabet soup of techniques including: IVF, GIFT, ZIFT, ICSI, and FET.

In its simplest term, IVF is simply the uniting of egg and sperm in vitro (in the lab). Subsequently the embryos are transferred into the uterus through the cervix and pregnancy is allowed to begin. IVF was the first of the ART techniques to be developed. The first birth was in 1978 in England. The procedure was pioneered by a Gynaecologist and a Ph.D. (Drs. Steptoe and Edwards). Next came GIFT, which stands for gamete (egg and sperm) intrafallopian transfer. This procedure requires laparoscopy, which is a small incision surgery and requires a general anaesthetic. With existing technology, pregnancy rates are similar with IVF and GIFT. Since IVF does not require surgery, it has supplanted GIFT.

ZIFT involves IVF and then a laparoscopic surgical procedure to transfer the embryos into the fallopian tube. Since transferring embryos through the cervix with IVF gives the same pregnancy rate as ZIFT, and is nonsurgical, IVF has also supplanted GIFT.

THE IVF CYCLE

As the years have passed, IVF has become the dominant ART technology due to its simplicity, efficacy and lack of invasiveness. A typical IVF cycle begins with shutting down the ovaries. This is done with a medication known as a GnRH agonist. The most common drug such used is Lupron. Lupron is given for approximately two weeks after which the ovaries are shut down temporarily. The next phase involves stimulation of the ovaries with potent ovulation medications such as Pergonal. For a full description of these agents go to the page on ovulation medication. These injections are given for approximately 10 days.

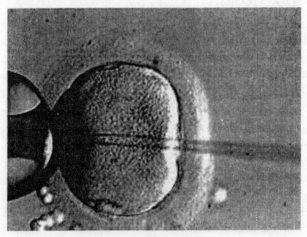

Fig: 4.1: In Vitro Fertilisation

When the eggs are ready for harvesting, a final step is to give hCG to induce final maturation. The eggs are then harvested by a process called ultrasound guided vaginal retrieval. Under heavy sedation, and with ultrasound guidance, a thin needle is passed a short distance into the ovaries and the eggs are suctioned from the follicles. Typically 5-15 eggs are collected.

Typically the eggs are fertilised by adding approximately 100,000 motile sperm to each egg. If the sperm will not fertilise the eggs naturally we can perform intracytoplasmic sperm injection (ICSI). This procedure involves puncturing the egg directly under a microscope and injecting one sperm in the egg.

The day following retrieval, we can document fertilisation under the microscope. We then observe the embryos for 3-6 days. The current trend is to observe longer. Typically 3-4 embryos are then placed in a catheter and transferred through the cervix into the uterus. This is a simple procedure much like a Pap smear. At the present time, embryos can be transferred either 3 or six days following retrieval. A 3-day embryo is usually at the 6-8-cell stage.

Two weeks later a pregnancy test can be obtained. Two weeks after the pregnancy test, an ultrasound can be performed and the fetal hear beat can be seen. If more embryos were generated than can be replaced, freezing (cryopreservation) can save these additional embryos. Frozen embryos can be stored for future replacement at much lower cost than the original IVF cycle.

As the years have passed, IVF has improved greatly. Today it is arguably the most effective technique to treat infertility when compared with others on a month by month basis. IVF has created a lot of controversy also. First, it is expensive. An IVF cycle can cost $6,000 to $7,000. It may not work on the first cycle. Multiple pregnancies can result. The truth is that it is a powerful technology and must be used carefully. Some patients may have very high odds of success: 45-60 per cent chance per attempt. Others may due to their situation have only a 20 per cent chance of success.

The multiple pregnancy risk varies with age. Younger patients need fewer embryos to be replaced, and older patients need more. The worst thing that has happened with IVF is the various centers entering into a race to see who can get "the best statistics". This has encouraged centers to transfer high numbers of embryos to get the statistics while accepting too high a risk of multiple pregnancy.

Also in order to get the best statistics, some patients will be refused care in order to "protect the statistics".

f) cells, trigger a local or widespread inflammatory response, and retain the memory of the offending organism to repel it again if it should ever return. Like any finely-tuned machine, however, the system can break down and leave us open to the threat of infection, or, conversely, turn against our own healthy tissues, as occurs in such diseases as rheumatoid arthritis or lupus.

The immune system also plays an important role in human reproduction. Inflammatory cells and their secretory products are involved in the processes of ovulation and preparation of the endometrium for implantation of a fertilized egg. Dysfunction of the immune system can interfere with the normal reproductive processes and result in infertility. It has been estimated that an immune factor may be involved in up to 20 per cent of couples with otherwise unexplained infertility. Although many of these associations with infertility remain unproven, there is solid scientific evidence to implicate the formation of antibodies against sperm as an important infertility factor.

CELL CULTURE

Cell cultures may contain following three types of cells:
1. Stem cells
2. Precursor cells
3. Differentiated cells

Stem cells are undifferentiated cells, which can differentiate under correct inducing conditions into several kinds of cells; different kinds of stem cells differ markedly in terms of the kinds of the cells in which they will differentiate into. Precursor cells are derived from stem cells, are committed to differentiation, but are not yet differentiated; these cells retain the capacity for proliferation. In contrast Differentiated cells usually do not have the capacity to divide. some cell cultures, e.g. epidermal keratinocyte cultures, contain all the

3 types of cells. In such cultures stem cells constantly provide, new cells which develop into precursor cells, which further proliferate and mature into differentiated cell types. Thus stem cells are necessary for maintenance of such cultures, which by their nature are heterogenous.

On the other hand, other cell cultures, e.g., fibroblast cultures, contain a more or less uniform population of dividing cells at low cell densities(<10^4 cells/cm^2), but at high cell densities(10^5 cells/ cm^2) are uniformly composed of non-proliferating differentiated cells. The cells begin to proliferate once the cell density is approximately reduced.

Differentiation and cell proliferation are affected by, in addition to cell density, factors like serum, Ca2+ ion, hormones, cell to cell and cell to matrix interactions etc. Generally, cell proliferations promoted by low cell density, low Ca2+ ions (100-600 M), and high growth factor levels, While differentiation is promoted by the exact opposite conditions and by the presence of differentiation inducing factors, e.g. cortisone, nerve growth factor, etc. The proportion of stem, Precursor and differentiated cells are markedly affected by the source tissue used for obtaining the cultures. For example cultures derived from embryos and those derived from even adult tissues where continuous cell renewal occurs naturally, e.g., intestinal epithelium, haemopoietic cells etc, stem cells are likely to be more frequent than in other cultures.

In contrast, cell cultures from tissues where renewal occurs only under stress, e.g., fibroblast, muscle, etc, may contain only precursor cells.

THE GROWTH OF CELL CULTURE

1. Monolayer or as
2. Suspension cultures.

Propagation in suspension cultures is limited to haemopoietic cell lines, ascites tumours and transformed cells. Transformed cells are those cells that have been phenotypically modified during in vitro culture to become anchor independent and are able to grow in layers of several cell thick, as against monolayer growth of non transformed cells. Therefore, cells in culture need to surface or substrate to adhere to so that they are able to proliferate. In contrast, cells that are unable to adhere to a substrate are unable to divide, i.e., their growth is anchorage dependent.

ORGAN CULTURE

In vitro culture and growth of organs or parts thereof in which their various tissue components, eg parenchyma and stroma, are preserved both in terms of their structure and functions so that the culture organs resemble closely the concerned organs in vivo is called organ culture

FEATURES

- New growth occurs in form of differentiated structures. e.g.- glandular structure is retained in glands
- Cultured organ retains its physiological features. e.g.- hormone dependent organ continue to be dependent.
- Morphogenesis in cultured fetal tissues ia more or less comparable to that in vivo
- Outgrowth of isolated cells from periphery of explants can be minimised by manipulating the culture conditions

ADVANTAGES

- explant remain comparable to the in vivo organs both in structure and function;
- developmental of fetal organ is comparable;
- it provides information on the pattern s of growth, differentiation, and development of an organ;
- in some cases, it can replace whole animals in experimentation as the results from are easier to interpret.

TECHNIQUES OF ORGAN CULTURE

Plasma clot method: The explant is culture on the surface of a clot consisting of chick (or other) plasma and chick embryo extract contained in a watch glass. The watch glass may or may not be closed with a glass lid and sealed with paraffin wax. This has been the classical technique for studying the morphogenesis in embryonic organ rudiments. It has been also modified to study the action of hormones, vitamins carcinogens etc on the adult mammalian tissues.

Raft method: The explant is placed on to a raft of lens paper or rayon acetate, which is floated on a serum in a watch glass. Rayon acetate rafts were made to float on the serum by treating their 4 corners by silicone. Similarly, the floatability of lens paper is enhanced by treating with silicone. on each raft 4 or more explants are placed. In a combination of raft and clot techniques, the explant are first placed on a suitable raft, which is then kept on a plasma clot. This modification makes media changes easy and prevents the sinking into the liquefied plasma.

Agar gel method: The medium (consisting of suitable salt solution, serum, chick embryo extract or a mixture of certain amino acids and vitamins) is gelled with 1 per cent agar. This method avoids the immersion of explants into the medium and permits the use of defined media. generally the explants need to be subculutured on fresh agar gels every 5-7 days. The agar gels are generally kept in embryological watch glasses and sealed with paraffin wax. This method is used to study many developmental aspects of normal organs as well as tumours.

Grid Method: Devised by trowell in 1954, this method utilises 25 mm X 25 mm pieces of suitable wiremesh or perforated stainless steelsheet whose edges are bent to form 4 legs of about 4 mm height. Skeletal tissues are generally placed directly on the grid but softer tissues like glandsor skin are first placed on rafts, which are then kept on the grids. the grid themselves are then placed on a culture chamber filled with fluid medium upto the grid; the chamber is supplied with a mixture of o_2 and co_2 to meet the high o_2 requirement of adult mammalian organs.

5

Breeding of Transgenic Animals

INTRODUCTION

Livestock Animals are a major source of food, milk, wool, leather and many other byproducts. Naturally has always been an interest in improving the efficiency of livestock production. Until recently selective breeding was the only way to enhance genetic features of animals. However, this method is time consuming and can introduce limited variation. New gene transfer techniques and introduction of cloned genes into fertilized eggs, successful implantation of modified eggs into receptive female and getting progeny carrying cloned gene has resulted into development of new area of transgenic animals. Transgenic animal is a fertile animal that carries an introduced gene(s) in its germ line.

WHAT IS A TRANSGENIC ANIMAL?

There are various definitions for the term transgenic animal. The Federation of European Laboratory Animal Associations defines the term as an animal in which there has been a deliberate modification of its genome, the genetic make-up of an organism responsible for inherited characteristics.

The nucleus of all cells in every living organism contains genes made up of DNA. These genes store information that regulates how our bodies form and function. Genes can be altered artificially, so that some characteristics of an animal are changed. For example, an embryo can have an extra, Functioning gene from another source artificially introduced into it, or a gene introduced which can knock out the functioning of another particular gene in the embryo. Animals that have their DNA manipulated in this way are knows as transgenic animals.

The majority of transgenic animals produced so far are mice, the animal that pioneered the technology. The first successful transgenic animal was a mouse. A few years later, it was followed by rabbits, pigs, sheep, and cattle.

THE REASONS FOR TRANSGENIC ANIMAL'S PRODUCTION

The two most common reasons are:

- Some transgenic animals are produced for specific economic traits. For example, transgenic cattle were created to produce milk containing particular human proteins, which may help in the treatment of human emphysema.
- Other transgenic animals are produced as disease models (animals genetically manipulated to exhibit disease symptoms so that effective treatment can be studied). For example, Harvard scientists made a major scientific breakthrough when they received a U.S. patent (the company DuPont holds exclusive rights to its use) for a genetically engineered mouse, called OncoMouse® or the Harvard mouse, carrying a gene that promotes the development of various human cancers.

HOW ARE TRANSGENIC ANIMALS PRODUCED?

Since the discovery of the molecular structure of DNA by Watson and Crick in 1953, molecular biology research has gained momentum. Molecular biology technology combines techniques and expertise from biochemistry, genetics, cell biology, developmental biology, and microbiology.

Scientists can now produce transgenic animals because, since Watson and Crick's discovery, there have been breakthroughs in:

- recombinant DNA (artificially-produced DNA)
- genetic cloning
- analysis of gene expression (the process by which a gene gives rise to a protein)
- genomic mapping

The underlying principle in the production of transgenic animals is the introduction of a foreign gene or genes into an animal (the inserted genes are called transgenes). The foreign genes "must be transmitted through the germ line, so that every cell, including germ cells, of the animal contain the same modified genetic material."(Germ cells are cells whose function is to transmit genes to an organism's offspring.)

To date, there are three basic methods of producing transgenic animals

- DNA microinjection
- Retrovirus-mediated gene transfer
- Embryonic stem cell-mediated gene transfer

Gene transfer by microinjection is the predominant method used to produce transgenic farm animals. Since the insertion of DNA results in a random process, transgenic animals are mated to ensure that their offspring acquire the desired transgene. However, the success rate of producing transgenic animals individually by these methods is very low and it may be more efficient to use cloning techniques to increase their numbers. For example, gene transfer studies revealed that only 0.6 per cent of transgenic pigs were born with a desired gene after 7,000 eggs were injected with a specific transgene.

1. DNA Microinjection

The mouse was the first animal to undergo successful gene transfer using DNA microinjection. This method involves:

- transfer of a desired gene construct (of a single gene or a combination of genes that are recombined and then cloned) from another member of the same species or from a different species into the pronucleus of a reproductive cell.

- the manipulated cell, which first must be cultured in vitro (in a lab, not in a live animal) to develop to a specific embryonic phase, is then transferred to the recipient female.

2. Retrovirus-Mediated Gene Transfer

A retrovirus is a virus that carries its genetic material in the form of RNA rather than DNA. This method involves:

- retroviruses used as vectors to transfer genetic material into the host cell, resulting in a chimera, an organism consisting of tissues or parts of diverse genetic constitution
- chimeras are inbred for as many as 20 generations until homozygous (carrying the desired transgene in every cell) transgenic offspring are born

The method was successfully used in 1974 when a simian virus was inserted into mice embryos, resulting in mice carrying this DNA.

HOW DO TRANSGENIC ANIMALS CONTRIBUTE TO HUMAN WELFARE?

The benefits of these animals to human welfare can be grouped into areas:

- Agriculture
- Medicine
- Industry

The examples below are not intended to be complete but only to provide a sampling of the benefits

1. Agricultural Applications

(a) Breeding

Farmers have always used selective breeding to produce animals that exhibit desired traits (e.g., increased milk production, high growth rate). Traditional breeding is a time-consuming, difficult task. When technology using molecular biology was developed, it became possible to develop traits in animals in a shorter time and with more precision. In addition, it offers the farmer an easy way to increase yields.

(b) Quality

Transgenic cows exist that produce more milk or milk with less lactose or cholesterol, pigs and cattle that have more meat on them, and sheep that grow more wool. In the past, farmers used growth hormones to spur the development of animals but this technique was problematic, especially since residue of the hormones remained in the animal product.

(c) Disease Resistance

Scientists are attempting to produce disease-resistant animals, such as influenza-resistant pigs, but a very limited number of genes are currently known to be responsible for resistance to diseases in farm animals.

2. Medical Applications

(a) Xenotransplantation

Patients die every year for lack of a replacement heart, liver, or kidney. For example, about 5,000 organs are needed each year in the United Kingdom alone. Transgenic pigs may provide the transplant organs needed to alleviate the shortfall. Currently, xenotransplantation is hampered by a pig protein that can cause donor rejection but research is underway to remove the pig protein and replace it with a human protein.

(b) Nutritional Supplements and Pharmaceuticals

Products such as insulin, growth hormone, and blood anti-clotting factors may soon be or have already been obtained from the milk of transgenic cows, sheep, or goats. Research is also underway to manufacture milk through transgenesis for treatment of debilitating diseases such as phenylketonuria (PKU), hereditary emphysema, and cystic fibrosis.

In 1997, the first transgenic cow, Rosie, produced human protein-enriched milk at 2.4 grams per litre. This transgenic milk is a more nutritionally balanced product than natural bovine milk and could be given to babies or the elderly with special nutritional or digestive needs.4,21,23 Rosie's milk contains the human gene alpha-lactalbumin.

(c) Human Gene Therapy

Human gene therapy involves adding a normal copy of a gene (transgene) to the genome of a person carrying defective copies of the gene. The potential for treatments for the 5,000 named genetic diseases is huge and transgenic animals could play a role. For example, the A. I. Virtanen Institute in Finland produced a calf with a gene that makes the substance that promotes the growth of red cells in humans.

3. Industrial Applications

In 2001, two scientists at Nexia Biotechnologies in Canada spliced spider genes into the cells of lactating goats. The goats began to manufacture silk along with their milk and secrete tiny silk strands from their body by the bucketful. By extracting polymer strands from the milk and weaving them into thread, the scientists can create a light, tough, flexible material that could be used in such applications as military uniforms, medical microsutures, and tennis racket strings.

Toxicity-sensitive transgenic animals have been produced for chemical safety testing. Microorganisms have been engineered to produce a wide variety of proteins, which in turn can produce enzymes that can speed up industrial chemical reactions.

6

Transgenesis and Gene Therapy

INTRODUCTION

Molecular Biology encompasses the isolation of individual genes or gene families which may be used in "gene therapy"-style approaches in livestock. Isolated genes can be used to enhance cellular processes that are known to impinge upon production traits (in quantitative or qualitative ways) or to add novel functions to cells and tissues. In the preceeding years, SARDI has used this method and sheep wool keratin genes to modify wool fibre properties in sheep. This approach, known as transgenesis, involves transferring genes of interest from the chromosomes of one individual into the chromosomes of another individual. The transferred genes are known as transgenes. In one method (pronuclear microinjection) a transgene is injected into a livestock embryo at the single-cell stage and becomes inserted into one of the chromosomes. When this cell divides, the transgene is copied with the rest of the chromosomal DNA such that as the embryo develops to full-term, each cell in the animal born has the transgene. In another method (somatic cell nuclear transfer, or "cloning", as used recently to produce the sheep Matilda) a gene is first transferred into the chromosomes of a cell in culture then that cell's nucleus is is used to replace the nucleus of an oocyte. During subsequent development in culture, then in utero and at term, this animal also carries the extra gene. In both methods, a transgenic animal is made. When single genes are transferred to embryos (more than one can be used) the genetic alteration to the resulting animal is quite precise; only a single gene is changed while all other genes in the chromosomes are generally unaltered. (The latter "cloning" methodology is preferred as it also allows the chromosomal insertion point of the transgene to be controlled.) In this way, the genetics of a valuable animal may be preserved and added to in a defined manner, so that the effects of transgenes and their expression can be accurately assessed. In the case of sheep wool, transgenesis provides an opportunity to develop wools with different protein compositions and properties that could be an advantage during processing or which might provide novel qualities to the end product. We are awaiting new funding to produce more transgenic sheep, which carry new transgenes designed to modify wool properties and to promote wool growth.

GENE THERAPY

Genes, which are carried on chromosomes, are the basic physical and functional units of heredity. Genes are specific sequences of bases that encode instructions on how to make proteins. Although genes get a lot of attention, it's the proteins that perform most life functions and even make up the majority of cellular structures. When genes are altered so that the encoded proteins are unable to carry out their normal functions, genetic disorders can result.

Gene therapy is a technique for correcting defective genes responsible for disease development. Researchers may use one of several approaches for correcting faulty genes:

- A normal gene may be inserted into a nonspecific location within the genome to replace a non-functional gene. This approach is most common;
- An abnormal gene could be swapped for a normal gene through homologous recombination;
- The abnormal gene could be repaired through selective reverse mutation, which returns the gene to its normal function;

- The regulation (the degree to which a gene is turned on or off) of a particular gene could be altered.

In most gene therapy studies, a "normal" gene is inserted into the genome to replace an "abnormal,"disease-causing gene. A carrier molecule called a vector must be used to deliver the therapeutic gene to the patient's target cells. Currently, the most common vector is a virus that has been genetically altered to carry normal human DNA. Viruses have evolved a way of encapsulating and delivering their genes to human cells in a pathogenic manner. Scientists have tried to take advantage of this capability and manipulate the virus genome to remove disease-causing genes and insert therapeutic genes.

Target cells such as the patient's liver or lung cells are infected with the viral vector. The vector then unloads its genetic material containing the therapeutic human gene into the target cell. The generation of a functional protein product from the therapeutic gene restores the target cell to a normal state. See a diagram depicting this process.

EMBRYONIC STEM CELL-MEDIATED TRANSGENESIS

In the last two decades, the advent of novel and very efficient techniques to manipulate the mouse genome, such as in mouse ES cells, has given us the capability of tailoring the mouse genes and genomes at will. We have reached the stage at which geneticists studying mice believe there is no limitation to creating phenocopies (or genocopies) of any mutations or chromosomal aberrations identified in human diseases. Embryonic stem cells are derived from an early (blastocyst-stage) embryo and can be maintained in culture as undifferentiated, pluripotent cells under the proper growth conditions. A broad spectrum of strategies has been designed to create genomic alterations in these cells. When the genetically altered ES cells are injected into a host blastocyst, or aggregated within a morula-stage embryo, they have the capacity to contribute to all tissues of the resultant chimeric mouse (*See fig. 6.1 on next page*). Most important, they can contribute to germ cells and transmit the genetic mutations in vivo, allowing development of established mouse lines in which the altered gene(s) are carried.

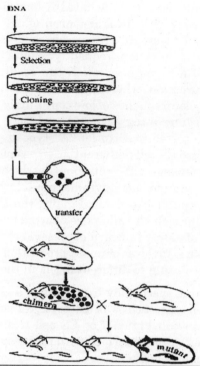

Fig. 6.1: Schematic drawing illustrating ES cell transgenesis that we have used to create the transgenic mouse GBM model. The transgene (for example, activated p21-ras) is transfected into ES cells under a GFAP promoter specifically expressed in astrocytes. The promoter is activated in ES cells (by retinoic acid) and using selection markers positive clones expressing the transgene are selected. These transfected ES cells undergo aggregations and are transferred into pseudopregnant mice to create chimeric embryos composed of cells arising from the transfected and wild-type ES cells, with the transgene only being expressed in a tissue-specific manner, such as occurs in astrocytes in the presence of the GFAP promoter. These chimeras are then crossed with normal mice to propagate stably the transgene in a germline fashion. This and other transgenic strategies have led to derivation of mouse astrocytoma models, which replicate the pathological and molecular profile found in human astrocytomas.

Embryonic stem cell-mediated transgenesis has several advantages over the standard pronuclear DNA injection routinely used to create transgenic models. Conventional pronuclear DNA injection frequently results in multiple-copy integration of a transgene, which can result in variation of transgene expression, whereas ES cell-mediated transgenesis provides a higher frequency of low-copy numbers or even single copy of transgene integration. In addition, transfected ES clones can be tested in vitro for cell type-specific expression by using in vitro ES cells-differentiation assays, as exemplified by astrocytic lineage differentiation in which retinoic acid is used in the mouse astrocytoma models derived in our laboratory (unpublished data). Finally, the ES cell-mediated transgenic approach also allows us to avoid the problem, found in a number of cases, in which expression of the transgene is lethal, because it involves chimera production with transgenic ES cells contributing to different levels in the animals.

To examine specifically the effects of a genetic alteration in a certain cell type, such as an astrocyte, cell-specific promoters can be used in conventional or ES cell transgenesis. In the nervous system, tumors formed with tissue-specific promoters in transgenic mice include:

1. pineoblastomas (tryptophan hydroxylase promoter);
2. primitive neuroectodermal tumours (tyrosine hydroxylase promoter);
3. oligodendrogliomas (myelin basic promoter and S-100 promoter);
4. neuroblastoma and ganglioneuromas (dopamine—hydroxylase promoter); and
5. gonadotrophic hormone-releasing tumours leutinising hormone/follicle-stimulating hormone promoter).

TRANSGENIC MODELS OF ASTROCYTOMAS

The GFAP promoter has been used to express oncogenes specifically in astrocytes, such as the SV40 large T antigen and v-src. The GFAP-SV40 large T transgenic mice were shown to develop aggressive non-astrocytic brain tumours with hyperplasia of the choroid plexus. In GFAP-v-src transgenic mice abnormal nests of proliferating astrocytes are formed by 2 weeks of age, and these astrocytes later evolve into overt malignant astrocytomas in the brain and spinal cord in 14 per cent of mice by 65 weeks of age. Hemizygosity for p53 or retinoblastoma, achieved by crossing these mice with the respective p53 and retinoblastoma knockout mice, was not shown to increase the incidence or shorten the latency of astrocytic tumours in these GFAP-v-src mice. These transgenic astrocytomas have histological features similar to human GBMs, including expression of VEGF and angiogenic endothelial-specific receptors such as flt-1, flk-1, tie-1 and tie-2.

GFAP has been used to express relevant transgenes in astrocytes, by using the ES cell strategy outlined previously (Fig. 6.1). Based on our prior work in which we found that one of the main signaling pathways activated by PDGFRs and EGFRs overexpressed in human GBMs involves activation of Ras, we have created a mouse GBM model by overexpressing oncogenic Ras (V12Ras) in astrocytes using the GFAP promoter (unpublished data). Ninety per cent of GFAP-V12Ras transgenic mice developed GFAP-positive mutifocal astrocytomas within 2 to 4 months. These tumours demonstrated a high mitotic index, nuclear pleomorphism, infiltration, and increased vascularity with overexpression of VEGF, similar to characteristics of human malignant astrocytomas. Derivative astrocytoma cells obtained in these mice are tumourgenic when inoculated into naïve syngeneic or immuno compromised mice. Cytogenetic analysis revealed consistent clonal aneuploidies of several chromosomal regions syntenic with comparable loci altered in human astrocytomas. For example, the transgenic mouse astrocytoma cells harboured trisomy of mouse chromosome 10, an area that contains human chromosome 12q, which is the second most commonly amplified region in human GBMs. Direct assay conducted to determine protein expression revealed decreased expression of p16, p19/p14ARF, p53, and PTEN and overexpression of proteins such as EGFR, MDM2, and CDK4, which is similar to findings in human malignant astrocytomas.

In addition, these GFAP-V Ras transgenic mice have been shown to possess genetic cooperativity, as they develop malignant astrocytomas within 2 to 3 weeks when crossed with transgenic mice in which expression of EGFRvIII in astrocytes occurs. These mice were developed using a similar strategy with the GFAP promoter in ES cells but, by themselves, did not result in astrocytoma formation (unpublished results). Additional crosses with mice harbouring knockout of relevant astrocytoma genes such as p53 and retinoblastoma are ongoing and may modulate the occurrence of the GBMs in these mice. Furthermore, the mice and derived astrocytoma lines are responsive to biological therapies targeting the Ras pathway, such as with farnesyl transferase inhibitors, which have shown promise in a variety of human tumours including, as found in our own studies, astrocytomas. The molecular-pathological similarities with human GBMs, in addition to some of these early, promising studies, leads us to believe that this transgenic mouse GBM model may serve to increase further our knowledge of the molecular pathogenesis, as well as serve as an appropriate preclinical model of human malignant astrocytomas.

TRANSGENIC ASTROCYTOMA MODELS

Although the knockout mouse models associated with cancer predisposition syndromes discussed in the preceding sections have not yielded useful models of astrocytomas, several other knockout mouse models with disruption of the genes that are lost in sporadic astrocytomas are of interest. Mice lacking p16, a CDK inhibitor that is found on chromosome 9p that is involved in the retinoblastoma-regulated cell cycle pathway, and that is commonly mutated in human astrocytoma specimens and derived cell lines, develop a variety of malignancies but not astrocytomas.However, these mice provide a useful background on which additional genetic alterations associated with astrocytomas and GBMs, such as overexpression of EGFRvIII transduced by retroviral injection, leads to development of astrocytoma-like lesions. Crossbreeding experiments with these p16 (-/-) mice with other genetically altered mice are ongoing and may yield additional astrocytoma models. A second gene of interest on chromosome

9p is p14ARF , which is also commonly mutated in astrocytomas, and is involved indirectly in the p53 pathway by its regulation of MDM2. A small proportion of p14ARF knockout mice develop glioma-like lesions presumably due to increasing MDM2 levels that lead to sequestration and secondary inhibition of p53 function The PTEN/MMAC1 on chromosome 10q, a dual specific and lipid phosphatase that is lost in the majority of GBMs, has also been targeted in mice. The PTEN null (-/-) mice are embryonically lethal, whereas heterozygous (+/-) mice develop a plethora of tumours but not astrocytomas. Current studies in which we use a variety of strategies, are underway to knockout PTEN/MMAC1 specifically in astrocytes, to avoid the embryonic lethality. These knockout models of astrocytoma-specific genetic alteration have not resulted in useful models of astrocytomas; however, they may augment and complement strategies in which the mouse genome has been altered for a gain of function in genetic aberration associated with astrocytomas, such as those used in our laboratory with ES cells.

DIFFERENT TYPES OF VIRUSES USED AS GENE THERAPY VECTORS

- *Retroviruses:* A class of viruses that can create double-stranded DNA copies of their RNA genomes. These copies of its genome can be integrated into the chromosomes of host cells. Human immunodeficiency virus (HIV) is a retrovirus.

- *Adenoviruses:* A class of viruses with double-stranded DNA genomes that cause respiratory, intestinal, and eye infections in humans. The virus that causes the common cold is an adenovirus.

- *Adeno-associated viruses:* A class of small, single-stranded DNA viruses that can insert their genetic material at a specific site on chromosome 19.

- *Herpes simplex viruses:* A class of double-stranded DNA viruses that infect a particular cell type, neurons. Herpes simplex virus type 1 is a common human pathogen that causes cold sores.

Besides virus-mediated gene-delivery systems, there are several non-viral options for gene delivery. The simplest method is the direct introduction of therapeutic DNA into target cells. This approach is limited in its application because it can be used only with certain tissues and requires large amounts of DNA.

Another non-viral approach involves the creation of an artificial lipid sphere with an aqueous core. This liposome, which carries the therapeutic DNA, is capable of passing the DNA through the target cell's membrane.

Therapeutic DNA also can get inside target cells by chemically linking the DNA to a molecule that will bind to special cell receptors. Once bound to these receptors, the therapeutic DNA constructs are engulfed by the cell membrane and passed into the interior of the target cell. This delivery system tends to be less effective than other options.

Researchers also are experimenting with introducing a 47th (artificial human) chromosome into target cells. This chromosome would exist autonomously alongside the standard 46 —not affecting their workings or causing any mutations. It would be a large vector capable of carrying substantial amounts of genetic code, and scientists anticipate that, because of its construction and autonomy, the body's immune systems would not attack it. A problem with this potential method is the difficulty in delivering such a large molecule to the nucleus of a target cell.

FDA's Biological Response Modifiers Advisory Committee (BRMAC) met at the end of February 2007 to discuss possible measures that could allow a number of retroviral gene therapy trials for treatment of life-threatening diseases to proceed with appropriate safeguards. In April of 2007 the FDA eased the ban on gene therapy trials using retroviral vectors in blood stem cells.

What factors have kept gene therapy from becoming an effective treatment for genetic disease?

Short-lived nature of gene therapy: Before gene therapy can become a permanent cure for any condition, the therapeutic

DNA introduced into target cells must remain functional and the cells containing the therapeutic DNA must be long-lived and stable. Problems with integrating therapeutic DNA into the genome and the rapidly dividing nature of many cells prevent gene therapy from achieving any long-term benefits. Patients will have to undergo multiple rounds of gene therapy.

Immune response: Anytime a foreign object is introduced into human tissues, the immune system is designed to attack the invader. The risk of stimulating the immune system in a way that reduces gene therapy effectiveness is always a potential risk. Furthermore, the immune system's enhanced response to invaders it has seen before makes it difficult for gene therapy to be repeated in patients.

Problems with viral vectors: Viruses, while the carrier of choice in most gene therapy studies, present a variety of potential problems to the patient —toxicity, immune and inflammatory responses, and gene control and targeting issues. In addition, there is always the fear that the viral vector, once inside the patient, may recover its ability to cause disease.

Multigene disorders: Conditions or disorders that arise from mutations in a single gene are the best candidates for gene therapy. Unfortunately, some the most commonly occurring disorders, such as heart disease, high blood pressure, Alzheimer's disease, arthritis, and diabetes, are caused by the combined effects of variations in many genes. Multigene or multifactorial disorders such as these would be especially difficult to treat effectively using gene therapy. For more information on different types of genetic disease, see Genetic Disease Information.

THE RECENT DEVELOPMENTS IN GENE THERAPY RESEARCH

Results of world's first gene therapy for inherited blindness show sight improvement. On 28 April, 2008, UK researchers from the UCL Institute of Ophthalmology and Moorfields Eye Hospital NIHR Biomedical Research Centre have announced results from the world's first clinical trial to test a revolutionary gene therapy treatment for a type of inherited blindness. The results, published in the New England Journal of Medicine, show that the experimental treatment is safe and can improve sight. The findings are a landmark for gene therapy technology and could have a significant impact on future treatments for eye disease. Read Press Release.

Previous information on this trial (May 1, 2007): A team of British doctors from Moorfields Eye Hospital and University College in London conduct first human gene therapy trials to treat Leber's congenital amaurosis, a type of inherited childhood blindness caused by a single abnormal gene. The procedure has already been successful at restoring vision for dogs. This is the first trial to use gene therapy in an operation to treat blindness in humans.

A combination of two tumour suppressing genes delivered in lipid-based nanoparticles drastically reduces the number and size of human lung cancer tumours in mice during trials conducted by researchers from The University of Texas M. D. Anderson Cancer Center and the University of Texas Southwestern Medical Center.

Researchers at the National Cancer Institute (NCI), part of the National Institutes of Health, successfully reengineer immune cells, called lymphocytes, to target and attack cancer cells in patients with advanced metastatic melanoma. This is the first time that gene therapy is used to successfully treat cancer in humans.

Gene therapy is effectively used to treat two adult patients for a disease affecting non-lymphocytic white blood cells called myeloid cells. Myeloid disorders are common and include a variety of bone marrow failure syndromes, such as acute myeloid leukaemia. The study is the first to show that gene therapy can cure diseases of the myeloid system.

Gene Therapy cures deafness in guinea pigs. Each animal had been deafened by destruction of the hair cells in the cochlea that translate sound vibrations into nerve signals. A gene, called Atoh1, which stimulates the hair cells' growth, was delivered to the cochlea by an adenovirus. The genes triggered re-growth of the hair cells and many of the animals regained up to 80 per cent of their original hearing thresholds. This study, which many pave the way to human trials of the gene, is the first to show that gene therapy can repair deafness in animals.

University of California, Los Angeles, research team gets genes into the brain using liposomes coated in a polymer call polyethylene glycol (PEG). The transfer of genes into the brain is a significant achievement because viral vectors are too big to get across the "blood-brain barrier." This method has potential for treating Parkinson's disease.

RNA interference or gene silencing may be a new way to treat Huntington's. Short pieces of double-stranded RNA (short, interfering RNAs or siRNAs) are used by cells to degrade RNA of a particular sequence. If a siRNA is designed to match the RNA copied from a faulty gene, then the abnormal protein product of that gene will not be produced.

New gene therapy approach repairs errors in messenger RNA derived from defective genes. Technique has potential to treat the blood disorder thalassaemia, cystic fibrosis, and some cancers.

Gene therapy for treating children with X-SCID (sever combined immunodeficiency) or the "bubble boy" disease is stopped in France when the treatment causes leukaemia in one of the patients.

Researchers at Case Western Reserve University and Copernicus Therapeutics are able to create tiny liposomes 25 nanometers across that can carry therapeutic DNA through pores in the nuclear membrane.

7
Animal Cloning

INTRODUCTION

Animal Cloning is the process by which an entire organism is reproduced from a single cell taken from the parent organism and in a genetically identical manner. This means the cloned animal is an exact duplicate in every way of its parent; it has the same exact DNA.

Cloning happens quite frequently in nature. A sexual reproduction in certain organisms and the development of twins from a single fertilised egg are both instances of Cloning.

With the advancement of biological technology, it is now possible to artificially recreate the process of Animal Cloning.

The possibility of human cloning, raised when Scottish scientists at Roslin Institute created the much-celebrated sheep "Dolly" (Nature 385, 810-13, 1997), aroused worldwide interest and concern because of its scientific and ethical implications. The feat, cited by Science magazine as the breakthrough of 1997, also generated uncertainty over the meaning of "cloning" —an umbrella term traditionally used by scientists to describe different processes for duplicating biological material.

When the media report on cloning in the news, they are usually talking about only one type called reproductive cloning. There are different types of cloning, however, and cloning

Animal Cloning

technologies can be used for other purposes besides producing the genetic twin of another organism. A basic understanding of the different types of cloning is key to taking an informed stance on current public policy issues and making the best possible personal decisions.

Dolly, the first mammal to be cloned from adult DNA, was put down by lethal injection Feb. 14, 2003. Prior to her death, Dolly had been suffering from lung cancer and crippling arthritis. Although most Finn Dorset sheep live to be 11 to 12 years of age, postmortem examination of Dolly seemed to indicate that, other than her cancer and arthritis, she appeared to be quite normal. The unnamed sheep from which Dolly was cloned had died several years prior to her creation. Dolly was a mother to six lambs, bred the old-fashioned way.

DEVELOPMENT OF ANIMAL CLONING IN THE LAB

Scientists have been attempting to clone animals for a very long time. Many of the early attempts came to nothing. The first fairly successful results in animal cloning were seen when tadpoles were cloned from frog embryonic cells. This was done by the process of nuclear transfer. The tadpoles so created did not survive to grown into mature frogs, but it was a major breakthrough nevertheless.

After this, using the process of nuclear transfer on embryonic cells, scientists managed to produce clones of mammals. Again the cloned animals did not live very long. The first successful instance of animal cloning was that of Dolly the Sheep, who not only lived but went on to reproduce herself and naturally. Dolly was created by Ian Wilmut and his team at the Roslyn Institute in Edinburgh, Scotland, in 1997. Unlike previous instances, she was not created out of a developing embryonic cell, but from a developed mammary gland cell taken from a full-grown sheep.

Since then Scientists have been successful in producing a variety of other animals like rats, cats, horses, bullocks, pigs, deer, etc. You can even clone human beings now and that has given rise to a whole new ethical debate. Is it okay to duplicate nature to this extent? Is it okay to produce human clones? What would that do to the fabric of our society?

THE PROCESS OF ANIMAL CLONING

Initial attempts at artificially induced Animal Cloning were done using developing embryonic cells. The DNA nucleus was extracted from an embryonic cell and implanted into an unfertilised egg, from which the existing nucleus had already been removed. The process of fertilisation was simulated by giving an electric shock or by some chemical treatment method. The cells that developed from this artificially induced union were then implanted into host mothers. The cloned animal that resulted had a genetic make-up exactly identical to the genetic make-up of the original cell.

Since Dolly, of course, it is now possible to create clones from non-embryonic cells.

Now animal cloning can be done both for reproductive and non-reproductive or therapeutic purposes. In the second case, cloning is done to produce stem cells or other such cells that can be used for therapeutic purposes, for example, for healing or recreating damaged organs; the intention is not to duplicate the whole organism.

THE TYPES OF CLONING TECHNOLOGIES
(a) Recombinant DNA Technology or DNA Cloning

The terms "recombinant DNA technology," "DNA cloning," "molecular cloning," and "gene cloning" all refer to the same process: the transfer of a DNA fragment of interest from one organism to a self-replicating genetic element such as a bacterial plasmid. The DNA of interest can then be propagated in a foreign host cell. This technology has been around since the 1970s, and it has become a common practice in molecular biology labs today.

Scientists studying a particular gene often use bacterial plasmids to generate multiple copies of the same gene. Plasmids are self-replicating extra-chromosomal circular DNA molecules, distinct from the normal bacterial genome (*See fig. 7.1 on next page*). Plasmids and other types of cloning vectors were used by Human Genome Project researchers to copy genes and other pieces of chromosomes to generate enough identical material for further study.

To "clone a gene," a DNA fragment containing the gene of interest is isolated from chromosomal DNA using restriction enzymes and then united with a plasmid that has been cut with the same restriction enzymes. When the fragment of chromosomal DNA is joined with its cloning vector in the lab, it is called a "recombinant DNA molecule." Following introduction into suitable host cells, the recombinant DNA can then be reproduced along with the host cell DNA. See a diagram depicting this process.

Plasmids can carry up to 20,000 bp of foreign DNA. Besides bacterial plasmids, some other cloning vectors include viruses, bacteria artificial chromosomes (BACs), and yeast artificial chromosomes (YACs). Cosmids are artificially constructed cloning vectors that carry up to 45 kb of foreign DNA and can be packaged in lambda phage particles for infection into E. coli cells. BACs utilise the naturally occurring F-factor plasmid found in E. coli to carry 100 to 300-kb DNA inserts. A YAC is a functional chromosome derived from yeast that can carry up to 1 MB of foreign DNA. Bacteria are most often used as the host cells for recombinant DNA molecules, but yeast and mammalian cells also are used.

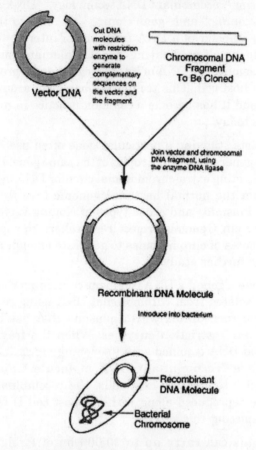

Fig: 7.1: **Cloning DNA in Plasmids.**

By fragmenting DNA of any origin (human, animal, or plant) and inserting it in the DNA of rapidly reproducing foreign cells, billions of copies of a single gene or DNA segment can be produced in a very short time. DNA to be cloned is inserted into a plasmid (a small, self-replicating circular molecule of DNA) that is separate from chromosomal DNA. When the recombinant plasmid is introduced into bacteria, the newly inserted segment will be replicated along with the rest of the plasmid.

(b) Reproductive Cloning

Reproductive cloning is a technology used to generate an animal that has the same nuclear DNA as another currently or previously existing animal. Dolly was created by reproductive cloning technology. In a process called "somatic cell nuclear transfer" (SCNT), scientists transfer genetic material from the nucleus of a donor adult cell to an egg whose nucleus, and thus its genetic material, has been removed. The reconstructed egg containing the DNA from a donor cell must be treated with chemicals or electric current in order to stimulate cell division. Once the cloned embryo reaches a suitable stage, it is transferred to the uterus of a female host where it continues to develop until birth.

Dolly or any other animal created using nuclear transfer technology is not truly an identical clone of the donor animal. Only the clone's chromosomal or nuclear DNA is the same as the donor. Some of the clone's genetic materials come from the mitochondria in the cytoplasm of the enucleated egg. Mitochondria, which are organelles that serve as power sources to the cell, contain their own short segments of DNA. Acquired mutations in mitochondrial DNA are believed to play an important role in the aging process.

Dolly's success is truly remarkable because it proved that the genetic material from a specialised adult cell, such as an udder cell programmed to express only those genes needed by udder cells, could be reprogrammed to generate an entire new organism. Before this demonstration, scientists believed that once a cell became specialised as a liver, heart, udder, bone, or any other type of cell, the change was permanent and other unneeded genes in the cell would become inactive. Some scientists believe that errors or incompleteness in the reprogramming process cause the high rates of death, deformity, and disability observed among animal clones.

(c) Therapeutic Cloning

Therapeutic cloning, also called "embryo cloning," is the production of human embryos for use in research. The goal of this process is not to create cloned human beings, but rather to harvest stem cells that can be used to study human development and to treat disease. Stem cells are important to biomedical researchers because they can be used to generate virtually any type of specialised cell in the human body. Stem cells are extracted from the egg after it has divided for 5 days. The egg at this stage of development is called a blastocyst. The extraction process destroys the embryo, which raises a variety of ethical concerns. Many researchers hope that one day stem cells can be used to serve as replacement cells to treat heart disease, Alzheimer's, cancer, and other diseases.

In November 2001, scientists from Advanced Cell Technologies (ACT), a biotechnology company in Massachusetts, announced that they had cloned the first human embryos for the purpose of advancing therapeutic research. To do this, they collected eggs from women's ovaries and then removed the genetic material from these eggs with a needle less than 2/10,000th of an inch wide. A skin cell was inserted inside the enucleated egg to serve as a new nucleus. The egg began to divide after it was stimulated with a chemical called ionomycin. The results were limited in success. Although this process was carried out with eight eggs, only three began dividing, and only one was able to divide into six cells before stopping.

(d) How can cloning technologies be used?

Recombinant DNA technology is important for learning about other related technologies, such as gene therapy, genetic engineering of organisms, and sequencing genomes. Gene therapy can be used to treat certain genetic conditions by introducing virus vectors that carry corrected copies of faulty genes into the cells of a host organism. Genes from different organisms that improve taste and nutritional value or provide resistance to particular types of disease can be used to genetically engineer food crops. See Genetically Modified Foods and Organisms for more information. With genome sequencing, fragments of chromosomal DNA must be inserted into different cloning vectors to generate fragments of an appropriate size for sequencing. (*See Fig. 7.2 on next page*).

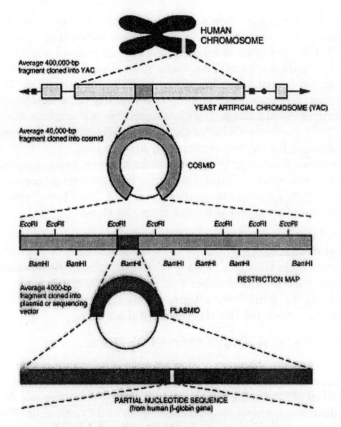

Fig. 7.2. Constructing Clones for Sequencing.

Cloned DNA molecules must be made progressively smaller and the fragments subcloned into new vectors to obtain fragments small enough for use with current sequencing technology. Sequencing results are compiled to provide longer stretches of sequence across a chromosome.

If the low success rates can be improved (Dolly was only one success out of 276 tries), reproductive cloning can be used to develop efficient ways to reliably reproduce animals with special qualities. For example, drug-producing animals or animals that have been genetically altered to serve as models for studying human disease could be mass produced.

Reproductive cloning also could be used to repopulate endangered animals or animals that are difficult to breed. In 2001, the first clone of an endangered wild animal was born, a wild ox called a gaur. The young gaur died from an infection about 48 hours after its birth. In 2001, scientists in Italy reported the successful cloning of a healthy baby mouflon, an endangered wild sheep. The cloned mouflon is living at a wildlife center in Sardinia. Other endangered species that are potential candidates for cloning include the African bongo antelope, the Sumatran tiger, and the giant panda. Cloning extinct animals presents a much greater challenge to scientists because the egg and the surrogate needed to create the cloned embryo would be of a species different from the clone.

Therapeutic cloning technology may some day be used in humans to produce whole organs from single cells or to produce healthy cells that can replace damaged cells in degenerative diseases such as Alzheimer's or Parkinson's. Much work still needs to be done before therapeutic cloning can become a realistic option for the treatment of disorders.

WHAT ANIMALS HAVE BEEN CLONED?

Scientists have been cloning animals for many years. In 1952, the first animal, a tadpole, was cloned. Before the creation of Dolly, the first mammal cloned from the cell of an adult animal, clones were created from embryonic cells. Since Dolly, researchers have cloned a number of large and small animals including sheep, goats, cows, mice, pigs, cats, rabbits, and a gaur. All these clones were created using nuclear transfer technology.

Hundreds of cloned animals exist today, but the number of different species is limited. Attempts at cloning certain species such as monkeys, chickens, horses, and dogs, have been unsuccessful. Some species may be more resistant to somatic cell nuclear transfer than others. The process of stripping the nucleus from an egg cell and replacing it with the nucleus of a donor cell is a traumatic one, and improvements in cloning technologies may be needed before many species can be cloned successfully.

CAN ORGANS BE CLONED FOR USE IN TRANSPLANTS?

Scientists hope that one day therapeutic cloning can be used to generate tissues and organs for transplants. To do this, DNA would be extracted from the person in need of a transplant and inserted into an enucleated egg. After the egg containing the patient's DNA starts to divide, embryonic stem cells that can be transformed into any type of tissue would be harvested. The stem cells would be used to generate an organ or tissue that is a genetic match to the recipient. In theory, the cloned organ could then be transplanted into the patient without the risk of tissue rejection. If organs could be generated from cloned human embryos, the need for organ donation could be significantly reduced.

Many challenges must be overcome before "cloned organ" transplants become reality. More effective technologies for creating human embryos, harvesting stem cells, and producing organs from stem cells would have to be developed. In 2001, scientists with the biotechnology company Advanced Cell Technology (ACT) reported that they had cloned the first human embryos; however, the only embryo to survive the cloning process stopped developing after dividing into six cells. In February 2002, scientists with the same biotech company reported that they had successfully transplanted kidney-like organs into cows. The team of researchers created a cloned cow embryo by removing the DNA from an egg cell and then injecting the DNA from the skin cell of the donor cow's ear. Since little is known about manipulating embryonic stem cells from cows, the scientists let the cloned embryos develop into fetuses. The scientists then harvested fetal tissue from the clones and transplanted it into the donor cow. In the three months of observation following the transplant, no sign of immune rejection was observed in the transplant recipient.

Another potential application of cloning to organ transplants is the creation of genetically modified pigs from which organs suitable for human transplants could be harvested. The transplant of organs and tissues from animals to humans is called xenotransplantation.

Why pigs? Primates would be a closer match genetically to humans, but they are more difficult to clone and have a much lower rate of reproduction. Of the animal species that have been cloned successfully, pig tissues and organs are more similar to those of humans. To create a "knock-out" pig, scientists must inactivate the genes that cause the human immune system to reject an implanted pig organ. The genes are knocked out in individual cells, which are then used to create clones from which organs can be harvested. In 2002, a British biotechnology company reported that it was the first to produce "double knock-out" pigs that have been genetically engineered to lack both copies of a gene involved in transplant rejection. More research is needed to study the transplantation of organs from "knock-out" pigs to other animals.

WHAT ARE THE RISKS OF CLONING?

Reproductive cloning is expensive and highly inefficient. More than 90 per cent of cloning attempts fail to produce viable offspring. More than 100 nuclear transfer procedures could be required to produce one viable clone. In addition to low success rates, cloned animals tend to have more compromised immune function and higher rates of infection, tumour growth, and other disorders. Japanese studies have shown that cloned mice live in poor health and die early. About a third of the cloned calves born alive have died young, and many of them were abnormally large. Many cloned animals have not lived long enough to generate good data about how clones age. Appearing healthy at a young age unfortunately is not a good indicator of long-term survival. Clones have been known to die mysteriously. For example, Australia's first cloned sheep appeared healthy and energetic on the day she died, and the results from her autopsy failed to determine a cause of death.

In 2002, researchers at the Whitehead Institute for Biomedical Research in Cambridge, Massachusetts, reported that the genomes of cloned mice are compromised. In analysing more than 10,000 liver and placenta cells of cloned mice, they discovered that about 4 per cent of genes function abnormally. The abnormalities do not arise from mutations in the genes but from changes in the normal activation or expression of certain genes.

Problems also may result from programming errors in the genetic material from a donor cell. When an embryo is created from the union of a sperm and an egg, the embryo receives copies of most genes from both parents. A process called "imprinting" chemically marks the DNA from the mother and father so that only one copy of a gene (either the maternal or paternal gene) is turned on. Defects in the genetic imprint of DNA from a single donor cell may lead to some of the developmental abnormalities of cloned embryos.

8

Genetic Engineering
(Pharm Animals)

INTRODUCTION

Genetic Engineering is a further way to exploit and abuse animals. This "new" way allows scientists to alter animals genetic make-up and even mix genetic material between species. What started out in the early 80s as a few tentative experiments has now exploded into a highly contentious free-for-all.

The transgenic animals are not only confined to the laboratory but they are being "created" with grotesque deformities. The major thrust of this work is aimed at creating tailor-made animals for research, engineering them to produce medical substances, and "improving" creatures for eating.

Joyce D'Silva, director of Compassion in World Farming has said, "I think (scientists) started out optimistic that they'd get faster growing super-animals, then they found out that they were often deformed and suffered greatly – surprise, surprise." As in the case of money, it is not difficult to see why this should happen.

There have been attempts to produce grazing pigs, pigs which produce appealing milk, and muscular chickens containing calf genes. More "successful" scientists have created turkeys which produce more eggs, and sheep containing mouse genes have been created to grow more wool. As for fish engineering, there are GE fish on sale and also there is plenty of work being done on them behind the scenes. Their main aim is create GE mega-fish which grow larger and faster. They have also tried putting flounder genes inside salmon to help them tolerate cold better and become disease resistant.

There are several noticeable risks associated with fish engineering. Changes in size and breeding patterns could effect all other creatures which feed on that fish, GE fish could displace other varieties hence further reducing the natural gene pool, and escapes from fish farms are so common now that in parts of Norway escaped fish outnumber wild ones at a ratio of 5:1. What else? Another risk that grows out of genetic engineering replicates is this tendency toward creation of genetic uniformity, the emergence of harmful recessives, and, of course, the greater susceptibility of organisms to devastation by pathogens, as has been shown to be the case in crops.

A further category of risks is environmental and ecological and is a direct result from "pharming" animals. The risk is associated with releasing GE animals into the environment, but with negative, unanticipated consequences. What animals have been released into the environment? Predator insects have been "designed" to perform biological pest control by preying on noxious insects . The more any organism differs from its parent stock, the more difficult it will be to predict its effect on the environment. Genetic engineers invariably work in the dark and virtually nothing is known about how genes integrate into chromosomes. If you disbelieve us, we only need to cite intensive farming practice techniques to show how far removed we are, as a species, from understanding the unique symbiosis that exists between Nature and her inhabitants.

For example, with GE practices one faulty action will compound several others creating a chain reaction that was not imagined at the initial point of departure. If an animal such as a GE predator insect is too "successful" at what he does then this could cause crowding out of other species, plants and animals, that one did not wish to affect. Similarly, "new" animals could start displaying traits in an environment setting that were not evident in the laboratory or other "controlled" situations. Consequently, each transgenic animal is unique and may express entirely different characteristics than first thought. Because of this lack of control, mutants may be produced.

There is also the question of "wastage" of pharm animals who didn't make the grade. Multi-national companies working with transgenic animals would like to recoup some of their costs by selling them for meat. It is already happening in Australia.

Compassion in World Farming has said:

"Of the 50,000-100,000 genes in farm animals, we know the identity and function of only 1-2 per cent. Making modifications to this genome is like playing with a chemistry set which has had all the labels removed. Except that in the case of gene transfer, the experimental materials are living, sentient creatures, capable of feeling the pain that is caused when the experiments inevitably go wrong."

How insulting it is that these "smart" scientists now call these defenseless, disposable animals in their care "pharm" animals. It is too late to ask what it is that could possibility be a limiting factor to these people. Preserving the identity and autonomous integrity of the animal perhaps? That has already been violated. Morality perhaps? Ethics? A sense of decent boundaries? Now that this work has began it is disquieting to realise that there is no limiting factor as it is aggressively encouraged and even funded by governments, including our own.

Scientists do little to educate the public about moral interests for two reasons. Most scientists are trained to ignore moral issues and secondly it is not in their best interest to warrant public scrutiny over their laboratory, behind-closed-doors practices.

The good news is unlike vivisection, if GE "fails", if there are any catastrophic outcomes of GE (especially of humans) then this is likely to eventuate in the imposition of severe restrictions on both research and its practical applications and in wholesale public rejection of biotechnology. Our choice as consumers, also, is not limited to collaboration or subjugation. We cannot blindly hope that a wiser council will prevail. Taxpayers have a right to know the truth, and not to be bamboozled with government and "industry" propaganda. The best thing we can do is lobby government and consciously avoid buying any products from a company which uses GE techniques in their manufacturing – don't forget to write and tell that Company what you are doing. If money really does talk then they will soon get the message.

GENETIC MODIFICATION OF FARM ANIMALS

The genetic modification of plants has become an issue of major public concern. Although the equivalent procedures in animals are much further away from commercial application in food production, they are used experimentally for several purposes.

Humans have been involved in genetic modification of animals, whether knowingly or not, since domestication began about 12,000 years ago. For most of this time genetic modification has been brought about simply by identifying and breeding from animals which best suited human needs for food, clothing, transport or draught power. The increase in the human population in the 18th century led to greater emphasis on 'selective breeding' (picking the best animals to be parents of the next generation) for increased output of animal products such as meat, milk, fibre and eggs.

Many new breeding technologies simply accelerate these conventional selection methods, without directly modifying the animals' genetic make-up. For example, the new and controversial technique of cloning has been used mainly to produce animals with identical copies of existing, naturally occurring, combinations of genes.

What is the purpose of direct genetic modification of animals?

The direct genetic modification of animals has already, or could be, practised for three main purposes:

1. *Scientific and medical research*

Genetically modified animals, especially mice, are widely used in scientific research into how animals function. They are also used in research to help understand and develop treatments for both human and animal diseases.

2. *Treatment of human disease*

Some human proteins used in the treatment of diseases (e.g. cystic fibrosis), are in very short supply. To overcome this problem, genes coding for the production of some human proteins have been transferred into sheep or cattle, enabling these proteins to be produced in the milk of the genetically modified animals. Some proteins produced in this way are now in the late stages of clinical trials. Although still at an experimental stage, the production of animals to provide organs for transplants in humans is another application of direct genetic modification. In this case the modification is aimed at altering the immune system of the animals, so that their organs appear to be of human origin, and are therefore not rejected after transplantation.

PRODUCTION OF MODIFIED FOOD-PRODUCING ANIMALS

Potentially, direct genetic modification could be used to enhance the productivity of farm animals, or alter their products e.g. by creating strains which grow faster, which show greater resistance to disease, or which produce novel proteins in their eggs or milk that are beneficial to human health. Such applications are being investigated experimentally in some countries, but the routine use of direct genetic modification in food-producing animals is unlikely in the short to medium term. This is largely because there are few known single genes that have a major effect on economically important characteristics (most characteristics are influenced by many genes). Also, most agricultural products have a relatively low value, and gene transfer methods are inefficient and expensive.

In most countries technologies for the direct genetic modification of animals are strictly controlled. In the UK these procedures are regulated under the Animals Scientific

Procedures Act (1986). This requires scientific and ethical justification for each procedure. There is also UK and EU legislation governing the 'contained use' and 'deliberate release' of genetically modified organisms, including animals. This legislation is designed to protect both human health and the environment.

How is direct genetic modification achieved?

Techniques that allow the direct modification of the genes of animals were first established about 20 years ago. There are two main ways in which the genetic makeup of animals can be modified directly:

By altering the expression of existing genes. For example, it is possible to prevent, or knockout, the normal expression of some existing genes. This allows investigations of the function of particular genes. For instance, knockout mice are being widely used in research on cystic fibrosis, breast cancer, colon and other cancers in humans. This approach has not been used to date in farm livestock.

By adding new (foreign) sequences of DNA. Physically inserting the DNA coding for a gene with a desired effect into the DNA of another animal, is termed gene transfer, and the animals receiving the foreign DNA are called transgenic animals. DNA is chemically identical across species, and the genetic codes for producing particular proteins are the same across species. This means that it is possible to transfer genes not only within species, but also between species, and sometimes even between different classes of organism. For instance, bacterial and viral DNA has been introduced to a range of food crops to confer insect and virus resistance.

There are four steps involved in gene transfer:

1. **Identification of a gene with a significant and desirable effect**

This is probably the major technical limitation to gene transfer in animals at the moment. Mammals have 50-100,000 genes, but for only a small fraction of these is the function and exact location on the chromosomes known. Since most characteristics of economic importance in farm animals are influenced by many genes, transfer of one or a few genes is only likely to be useful in a few special cases (e.g. adding a gene conferring disease resistance, or significantly altering food quality.)

2. Introducing the DNA coding for the desired gene

This is usually achieved, in mammals, by direct injection of several hundred copies of the foreign DNA sequence into the nucleus of an early stage embryo. In mice, but not in farm livestock, foreign DNA can also be introduced via stem cells. These are cells, such as those found in early embryos, which are capable of developing into any organ or tissue. Transgenic poultry have been created using retroviruses carrying copies of the DNA coding for a new gene. The nuclear transfer technique used recently to produce cloned sheep offers an important new route for the introduction of foreign DNA. Nuclear transfer involves adding a cell from an embryo, or from a cell line maintained in the laboratory, to an unfertilised egg from which the nucleus has been removed. An electric current causes the fusion of the introduced cell and the unfertilised egg, and the new embryo then develops as if newly fertilised. The initial applications of this technique, including the creation of 'Dolly', did not involve genetically modified cells. However, foreign DNA can be introduced to cell lines in the laboratory, and cells that have the desired genetic modification can then be used for nuclear transfer.

3. Regulating expression of the introduced gene

It is clearly vitally important that the right genes get switched on in the right tissues, at the right time. Some early attempts at gene transfer in animals produced undesirable side effects, because genes were introduced without appropriate regulation of expression.

4. Confirming transmission of the transferred gene to the next generation of animals

This is achieved by testing samples of DNA from the offspring, to confirm that the foreign DNA is present.

Genetic Engineering (Pharm Animals)

ETHICAL ISSUES

The direct genetic modification of plants used in food production has raised public concerns over food safety, environmental risks and socio-economic effects as well as intrinsic concerns about human intervention in nature. There are likely to be additional concerns over direct genetic modification in animals.

Most people accept the use of animals for a range of purposes including food production, providing that the animals are treated humanely. In 1994 the UK government established a committee to consider the ethical implications of the use of new breeding technologies in farm livestock (the 'Banner Committee'), and it has since accepted the major recommendations of that committee. The Banner Committee suggested that the humane use of animals respect three principles:

- some treatments of animals are so harmful that they should never be permitted under any circumstances;
- if a harmful treatment is permitted, the harm it causes must be justified by the good being sought;
- steps should be taken to minimise any harm which is justified by the second principle.

The first principle would exclude the use of technologies, which are regarded as intrinsically objectionable. (There are many new technologies which are not intrinsically objectionable, and which may offer benefits for society or for animals e.g. by improving resistance to disease. Conversely, conventional selection techniques sometimes result in changes which many people regard as objectionable e.g. the extreme breast development, and associated difficulties with natural mating, seen in some strains of turkey.)

Both the creation of genetically modified organisms, and breeding from them, are controlled procedures already in the UK, requiring a licence from the Secretary of State. Before a licence is granted, the likely 'adverse effects' on the animal have to be weighed against the likely benefits of the modification.

It seems appropriate that any analysis of the technical aspects of livestock breeding technologies is accompanied by an analysis of the ethical implications. The framework and recommendations of the Banner Committee provide an important foundation for this, and for a more open dialogue in an area of public concern. In the longer term, this should help to restore public confidence in farming and science, whilst allowing maximum benefit to be derived from those technologies which are considered acceptable.

9
Biotechnology for Animal Breeding

INTRODUCTION

Biotechnology in animal breeding includes AI and ET in practical breeding, as well as DNA analyses for breed characterisation. There are certainly merits for these techniques. AI was initially developed to reduce the incidence of venereal disease in animals, but now allows the use of superior male animals on a larger scale than possible with natural service. If breeding or AI centres are available, smallholders – who often prefer keeping female animals – may no longer have to keep entire males. ET makes it easier to introduce exotic animals into countries that have strict quarantine requirements.

There are, however, a number of prerequisites for successful use of AI. Farmers need to recognise whether an animal is in heat; semen and insemination technicians must be within easy reach; liquid nitrogen has to be available etc. Even where it seems to work well, there are some dangers. In most cases, the semen is from animals of potentially high production in meat (rapid growth) or milk. High production and resistance to environmental stress are antagonists, i.e. high-yielding animals are less resistant to disease, more prone to heat stress, require more water than indigenous breeds, and need good-quality feed to achieve their production potential. For a dairy cow that produces 6000 or 8000 l of milk per lactation, straw with a digestibility of 50–55 per cent is not good forage, whereas indigenous breeds that need only to survive as a savings account can manage with it.

AI can also lead to loss in biodiversity. AI bulls can become semen millionaires, which may be good for the breeders and the AI businesses, but has disastrous impact on genetic variability within a breed. For example: Holstein Friesian (HF) is the most widely kept dairy cow in the world – currently at least 50 million cows. The push for maximum production means that many bulls are closely related. This leads to an international uniformity among HF dairy cattle. ET will further accelerate this dangerous trend towards uniformity among dairy cattle.

These trends fly in the face of the need to conserve animal genetic diversity. The poorer livestock-keepers in World have to cope with a great variability in ecological conditions and need animals adapted to the local environments. Uniform animals cannot serve this purpose. Indigenous animal genetic resources are needed for that purpose. (Vilakate et al 2007)

There are exceptions to using biotechnology only for high-yielding breeds. The Nguni are known to be well adapted to a harsh environment, can cope with low-quality forage, are fairly tick resistant and tolerate a range of diseases. However, the number of donor animals does not reflect the variability of types found in the larger Nguni cattle population. And this use of biotechnology raises another difficult issue: the Nguni were originally selected by Zulu cattle breeders, but the South African Government successfully discouraged the Zulu from keeping Nguni, so white commercial farmers were the ones who conserved Nguni cattle and brought them back to fame. If the Nguni continue to be commercially successful, who should benefit?

The importance of indigenous animals was also highlighted in research in Ethiopia, which assigned economic values to the multi-functionality of goats, including their insurance value. The findings suggest that indigenous goats under improved management practices give higher total benefits to the poorer livestock-keeping households than do crossbred goats, even though the crossbred goats produce more milk.

AI and ET techniques are presently used mainly for high-yielding animals with high demands with respect to feed quality, sanitation and hygiene. If these large farms operate in a labour-intensive way, on-farm employment may contribute to poverty reduction, but this is not always the case. Another consideration is that breeding with the help of advanced biotechnology such as AI and ET takes the control over the breeding process out of the hands of livestock-keepers and puts it into the hands of commercial breeders/firms and breed societies. These modern breeding institutions cater primarily for large farms and generally disregard the specific requirements of small-scale farmers and pastoralists.

MOLECULAR GENETICS AND ANIMAL BREEDING

Molecular genetics is making an increasing contribution to animal breeding. This section provides the molecular bases of animal breeding and the study of structural and functional genomics, reviewing current and innovative methodologies and techniques. Great importance is given to the practical part of this section, which is developed in the laboratory and through bioinformatics exercises.

Learning outcomes are: (i) to update the molecular bases of animal genetics; (ii) to gain experience concerning the main methodologies and techniques used in the study of the genetic information; (iii) to acquire competence in the most relevant molecular analysis carried out in the laboratory, related to nucleic acids, sequencing and genotyping; and (iv) to develop skills in the use of bioinformatics tools.

MOLECULAR GENETICS—THE FUTURE FOR ANIMAL BREEDING

Genetics is often treated as the most up-to-date and exciting area of biology. That may be so, but farmers and veterinary surgeons have been practising effective genetics since the first time they chose to breed from one animal over another. With tools little more sophisticated than keen observation, common sense and a desire to improve their stock, farmers have created an array of different breeds, each with its own desirable characteristics.

Over 10,000 years ago early farmers selected those animals that were placid enough to control and eventually domesticate. Nowadays, we routinely select the cows with the best milk yield, milk composition or conformation to breed to AI bulls. In poultry production, only chickens with the highest food conversion efficiency are retained to breed subsequent commercial animals.

There are estimated to be approximately 800 cattle breeds worldwide, of which 270 are native to Europe. They have been developed by generations of farmers, for a variety of different breeding goals related to improving production traits and environmental adaptations. Desirable characteristics may be organised into a selection index, and expressed as a single value such as the relative breeding index (RBI) in dairy cows. This is a scientific way of combining a number of different heritable traits into a single measure which estimates the overall genetic merit of each potential breeding cow. Selected traits typically include yield, carcass and fitness scores.

What is new to farming and veterinary medicine is the realisation that the differences between breeds, and the particular traits of each breed, are controlled by genes. Why do Charolais tend to convert energy (from grass or silage) into muscle mass, while Holstein-Friesians are more likely to convert it into milk at the expense of muscle? The answer lies within each breed's genes. Simply speaking, genes are a set of instructions, contained within our cells, for controlling one or more characteristics e.g. growth or milk production. Genes are written in the language of DNA that reads like text on the page of a book. The differences in appearance between cattle breeds and the difference in their value as beef or milk breeds is ultimately controlled by difference in their DNA.

GENETIC VARIATION

All animals, including humans, have variation in these genes, which explains why some people are tall, while others are small. Sometimes a variant in a single gene causes a particular condition, but more often several to many genes are involved. It is these variations in genes that give rise to the variety of animal breeds we have today. The desirable or economically valuable version of the gene, such as the one that codes for extra muscle in the double-muscled Belgian Blue (*Fig. 9.1*), may be caused by a single mutation in the DNA.

Fig. 9.1: A Belgian Blue showing extra muscle mass known as 'double muscling'

The key is to identify those animals that carry the beneficial versions of these genes and breed from them. Production traits are already selected for, although we do not yet select at gene level. This should result in a more precise and faster breeding programme for desirable characteristics than one based on indirect methods such as milk recording and analysis or measurement of live-weight gain.

SELECTION FOR A STRONG IMMUNE RESPONSE

If we already routinely select animals on the basis of production traits, why not apply the same logic to the immune system? A functional immune system is critical to the health and normal development of all animals, including humans, cattle and chickens. All farmers and veterinary surgeons know the value of colostrum or beestings to a young calf or lamb.

Scientific analysis reveals that this early milk is full of molecules that provide protective immunity and explains the mechanism of its beneficial effects.

The immune system, from the physical barrier of the skin to complex cellular and molecular activities, protects us from the constant attack of microorganisms. AIDS patients, with compromised immune systems, are susceptible to many common infections that the rest of us throw off easily. In some cases, our immune system can itself cause rather than prevent disease: for example, the overactive immune system which is manifested in auto-immune diseases such as haemolytic anaemia.

Most people have an immune system that gets the balance right most of the time. However, even when the balance is right there are still differences between individuals, partly due to genetic variation.

Consider this the next time your friend catches the flu and, despite close contact, you don't. Some individuals' immune systems can better cope with an assault from a bacteria or virus. However, with a different microbe, the tables can be turned!

When a disease causing agent (protozoa, fungus, virus or bacterium) enters the body, say through cut skin, the pathogen comes in contact with blood cells. Receptors sticking out of these cells like antennae can recognise the pathogen and trigger an immune response (O'Neill, 2004). This trigger results in a number of biological effects such as increased white blood cells (to promote healing) and increased production of antiseptic agents. These antimicrobial peptides (AMPs) are a crucial weapon that we all produce on our skin and internally that aid the fight against microbes in general, particularly bacteria.

These peptides bind to bacteria, puncture holes in their membranes, and kill them—they are essentially personal antibiotics.

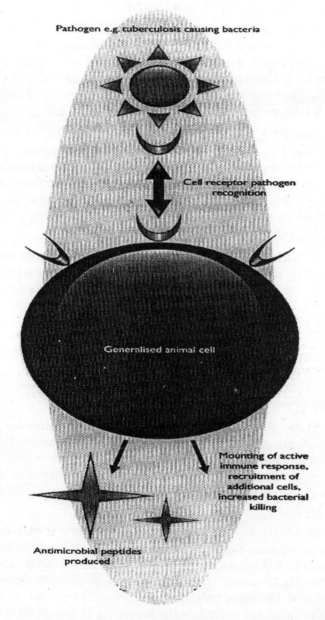

Fig. 9.2: Generalised diagram of the body's initial immune response

NEW SCIENTIFIC TECHNOLOGIES

Professor O'Farrelly's research group will employ a number of new state-of-the-art technologies to investigate the immune response to important diseases of humans and animals. Microarrays, for example, are essentially like an inverted hairbrush, where each strand represents a different gene in the genome of the animal under study.

The DNA from two samples (say one infected tissue sample and one healthy control tissue sample) is then added to the microarray. When a gene is present in equal amounts in both samples, the gene will show up yellow on the microarray. If the gene is expressed in the first (infected) sample only, it will light up red, whereas if the gene is only expressed in the second (healthy) sample, it will show up green.

In this way, the team can determine what genes are turned up, down, on or off during the course of the immune response to infection. It is these differentially expressed genes that are the most likely potential targets for new drugs, new treatments and new diagnostic techniques Application in animals Campylobacter is a bacteria that is a natural resident in the guts of chickens. Because it causes no harm, and may even have beneficial effects, it is known as a commensal. In humans, however, Campylobacter it is the single biggest cause of food poisoning worldwide.

State-of-the-art technologies will be used to uncover the mechanisms by which bacteria can exist happily and without ill-effect in one species while becoming pathogenic in another. This research will help us understand the mechanisms that allow the immune response to differentiate between 'good' and 'bad' bacteria.

Similarly, the technology will be important in bovine diseases such as mastitis, which costs the dairy industry a staggering $25 billion worldwide annually, and metritis which is probably the most common cause of reproductive disorder in the cow. The same bacterial species can be involved in both diseases and are increasingly becoming resistant to older generation antibiotics.

Marker-assisted selection Selecting an animal based on the genes it carries, will eventually become routine in large farm animal breeding. It is already used in pig breeding systems. this type of selection is known as marker-assisted selection (MAS). So, instead of looking for the BI of your cow, you might want to know what genes she carries that make her immune system stronger or her colostrum more protective for her calf.

THE FUTURE

Investment in scientific research has and can continue to provide numerous practical benefits. This research approach in particular has the potential to unlock the genetic diversity between individuals that accounts for variation in performance of both production traits such as milk yield, muscle mass and the immune system.

It is already known that certain breeds of animals can cope a lot better with disease than others. Scrapie is not known to occur in the Blackface breed of sheep whereas it does occur in the Texel and Suffolk breeds. Similarly, African humped cattle known as the Boran are susceptible to sleeping sickness, whereas the humpless N'Dama are tolerant to this disease. Susceptibility to these diseases are known to have a genetic basis. Eventually we will know enough to select superior animals for breeding so that we eliminate the common diseases of farm producing animals.

In the meantime, we can gain enough knowledge from these and similar research studies to identify new ways of diagnosing and treating these diseases. We can help tackle the problem of antibiotic over-usage and antibiotic resistant bacteria in food producing animals identified by the World Health Organisation (WHO) as a major future human health concern. If we reduced the disease burden by selecting animals with more robust immune systems, it would benefit the animal, producer and consumer.

10

Priorities in Animal Biotechnology

INTRODUCTION

Animal biotechnology worldwide has made rapid advances in animal production and health, of livestock and fish industries. The important areas include:

Animal cloning - transgenic animal for the introduction of important traits.

The species of animals involved include; poultry, cattle, sheep and fish

Development of rapid diagnostic kits

Development of recombinant and DNA vaccines

Improvement of reproductive biotechnology methods

Treatment of local feed resources with genetically engineered microbes

ANIMAL MODELS FOR THE STUDY OF HUMAN HEALTH

Indian scientists have been researching on various aspects of biotechnology as an alternative means of enhancing the economic production of livestock and fish. Expertise and laboratory facilities in biotechnology and allied disciplines were pooled and research projects coordinated and funded mainly by the research grants from the Ministry of Science, Technology and the Environment. Recent advances particularly under biotechnology were applied to develop rapid and cost effective diagnostic reagents, vaccines and other biologics. The areas include:

Manipulation of microorganism for better feed utilisation in cattle and poultry.

Genetic improvement of ruminants using new reproductive biotechnologies and recombinant DNA technology.

Development of diagnostic reagents and kits, for diseases of particular interest in fish, poultry and pig diseases.

Development of recombinant vectors for vaccine delivery.

Improvement of genetic resources for better commercial value in fish and wild-life species.

However, there are several issues and challenges faced by the group:

Lack of trained animal biotechnologists thus the pool of scientists is very limited.

Insufficient number of high technology equipments and these are scattered in various laboratories in different institutions.

Therefore, the establishment of well equipped laboratories are essential in bringing together scientists with common interests, in order to embark on high technologies.

THE LIVESTOCK INDUSTRIES

Many issues confront the livestock industries. The poultry and pig industries are highly dependent on imported feeds with negligible production of feed crops, and under-utilisation of agro-industrial by-products. Also, technologies for production of quality animal products (e.g. lean meat) are lacking.

The slow growth of the ruminant subsector may be because of the low base population of cattle, buffaloes, sheep and goats for rapid multiplication, limiting the ability for intensive selection for intensive selection for genetic improvement. Artificial insemination technology has not been fully utilised to multiply superior stock and produce crossbred animals because of poor conception rates. Efforts to utilise our own indigenous animal such as the Kedah Kelantan cattle have been minimal and disorganised.

Fish production can be intensified by good management over different functions and variables of the cultivated organisms, the stock and their environment. The concept of farming systems reflects the existence of relationships between resource variables; only certain systems are ecologically, biologically and technically viable.

To overcome the above constraints, biotechnology is rapidly acquiring a prominent place. Animal biotechnology research worldwide has made rapid advances in the production and health of livestock as well as fish. The application of new biotechnology methods in livestock industry, such as recombinant DNA techniques, cell culture, monoclonal antibody and others, have already generated a number of products for improving milk and meat production, animal health and food processing, and will definitely continue to do so. Products now emerging ranged from rapid diagnostic tests for contaminants in animal products to genetically engineered animals to produce high value pharmaceuticals. In the veterinary medical research, biotechnology was initially applied to vaccine development and diagnostics for early disease detection. Recently this novel technology has contributed toward efforts for detection. Recently this novel technology has contributed toward efforts for improvement of detoxification systems. The important areas of interest by animal researchers worldwide include:

Animal cloning: transgenic animals for the introduction of important traits. The species of animals involved ranged from cattle, sheep, poultry and fish.

DEVELOPMENT OF RAPID DIAGNOSTIC KITS

Development of recombinant and DNA vaccines improvement of reproductive biotechnology methods.

Treatment of unutilised local feed resources with genetically engineered microbes

Mass production of in vitro sexed embryos, using new reproductive biotechnologies and recombinant DNA technology

Development of recombinant vectors for vaccine delivery

Improvement of genetic resources for better commercial value in fish and wild-life species

Other basic research in animal biotechnology include isolation and characterisation of local organisms for the development of rapid and cost effective diagnostic reagents and kits, vaccines and other biologics.

Animal biotechnology research will involve not only animal researchers from various institutions, but also researchers from other related fields. The platform technologies such as genomics, transgenic technology, recombinant DNA, biosensor technology and bioprocessing will be utilised.

THE PRIORITY AREAS IDENTIFIED FOR ANIMAL BIOTECHNOLOGY

Genetic engineering of animals for improved production and quality Development of new generation vaccines and rapid diagnostic kits for important diseases

Novel vaccine and drug delivery systems

Development of cheap feedstuff from local resources

Improvement of reproductive technologies

Diseases of public health importance

Animal models for the study of human health.

To establish strong animal biotechnology expertise and teamwork through manpower training, collaborative and coordinated research and development programmes, towards a strong animal industry in India.

Priority areas in research and development for animal biotechnology are determined to meet the following goals:

To acquire leading edge technologies and establish strong research teams in essential areas to solve animal industry problems and to contribute towards improvement of human health;

To establish a strong livestock and fish industries for local and export markets through biotechnology findings and applications;

To ensure sufficient number of R&D critical mass are available through human resource development programmes;

To strengthen the linkages and interactions with international animal biotechnology experts in order to enhance research achievements.

The animal biotechnology R&D programme is set to achieve the following objectives:

To improve animal health management practices through the use of rapid, efficient and cheap diagnostic kits, recombinant vaccines and novel delivery systems, in order to reduce the heavy toll of morbidity and mortality on livestock and fish productivity;

To improve reproductive technologies towards increasing reproductive efficiency in ruminants, to ensure sufficient supply of meat in the country;

To develop feed production technologies which can improve the utilisation of relatively low cost by-products and other tropically adapted plant species.

11

Farm Animal Diversity

INTRODUCTION

The rapid spread of large-scale industrial livestock production focussed on a narrow range of breeds is the biggest threat to the world's farm animal diversity, according to a report presented to the Commission on Genetic Resources for Food and Agriculture.

Surging global demand for meat, milk and eggs has led to heavy reliance on high-output animals intensively bred to supply uniform products, according to The State of the World's Animal Genetic Resources for Food and Agriculture. The problem is compounded by the ease with which genetic material can now be moved around the world, says the report, which draws on information from 169 countries.

"In the next 40 years, the world's population will rise from today's 6.2 billion to 9 billion, with all the growth occurring in the developing countries," said FAO Assistant Director-General Alexander Müller in his address to the Commission. "We need to increase the resilience of our food supply, by maintaining and deploying the widest possible portfolio of genetic resources, which are vital and irreplaceable."

"Global warming is an additional threat to all genetic resources, increasing the pressure on biodiversity," Müller adds.

"Yet we need these genetic resources for the adaptation of agriculture to climate change."

MANAGEMENT OF ANIMAL GENITIC RESOURCES

"One livestock breed a month has become extinct over the past seven years, and time is running out for one-fifth of the world's breeds of cattle, goats, pigs, horses and poultry," says Müller. "This report, the first-ever global overview of livestock biodiversity and of the capacity within countries to manage their animal genetic resources, is a wake-up call to the world."

And this may only be a partial picture of the genetic erosion taking place, according to the report, as breed inventories are inadequate in many parts of the world. Moreover, among many of the most widely used high-output breeds of cattle, within-breed genetic diversity is being undermined by the use of a few highly popular sires for breeding.

"Effective management of animal genetic diversity is essential to global food security, sustainable development and the livelihoods of millions of people," says Irene Hoffmann, Chief of FAO's Animal Production Service.

"While sometimes less productive, many breeds at risk of extinction have unique characteristics, such as disease resistance or tolerance to climatic extremes, which future generations may need to draw on to cope with challenges such as climate change, emerging animal diseases and rising demand for specific livestock products," Hoffmann adds.

SOME BREEDS ARE MORE EQUAL

Well-adapted livestock have been an essential element of agricultural production systems for more than 10 000 years, especially important in harsh environments where crop farming is difficult or impossible.

Since the mid-twentieth century, a few high-performance breeds, usually of European descent – including Holstein-Friesian (by far the most widespread breed, reported in at least 128 countries and in all regions of the world) and Jersey cattle; Large White, Duroc and Landrace pigs; Saanen goats; and Rhode Island Red and Leghorn chickens – have spread throughout the world, often crowding out traditional breeds.

This progressive narrowing of genetic diversity is largely complete in Europe and North America and is now occurring in many developing countries, which have so far retained a large number of their indigenous breeds.

HOTSPOT OF BREED DIVERSITY LOSS

The developing world will be the hotspot of breed diversity loss in the twenty-first century, according to the report.

In Vietnam, for example, the percentage of indigenous sows declined from 72 per cent of the total population in 1994 to only 26 per cent in 2002. Of its 14 local breeds, five breeds are vulnerable, two in a critical state and three are facing extinction.

In Kenya, introduction of the Dorper sheep has caused the almost complete disappearance of pure-bred Red Maasai sheep.

Conservation programmes lacking: The crowding out of local breeds is set to accelerate in many developing countries, unless special provisions are made for their sustainable use and conservation by providing livestock keepers with adequate support, the report warns.

Effective management of animal genetic diversity requires resources – including well-trained personnel and adequate technical facilities – which many developing countries lack. According to the report, 48 per cent of the world's countries report no national in vivo conservation programmes, and 63 per cent report that they have no in vitro programmes, that is, the conservation of embryos, semen or other genetic material, with the potential to reconstitute live animals at a later date. Similarly, in many countries, structured breeding programmes are absent or ineffective.

"Support for developing countries and countries with economies in transition to characterise, conserve and utilise their livestock breeds will be necessary," says Clive Stannard of the Commission on Genetic Resources for Food and Agriculture. "Frameworks for wide access to animal genetic resources, and for equitable sharing of the benefits derived from them, need to be put in place, both at national and international levels."

PROTECTING OUR COMMON HERITAGE

The only international institution dealing with all genetic resources in agriculture, forestry and fisheries – experts from around the world are expected to endorse the findings of the report, which will be formally launched at the International Technical Conference on Animal Genetic Resources in Interlaken, Switzerland, in September 2007.

The Interlaken conference is expected to adopt a global plan of action to halt the loss of animal genetic resources and improve their sustainable use, development and conservation.

MAINTAINING GENETIC DIVERSITY OF LIVESTOCK

Many breeds at risk of extinction have unique characteristics and traits such as resistance to disease or adaptation to climatic extremes that could prove fundamental to the food security of future generations, FAO stressed.

Moreover, widely used breeds need to be managed more wisely. Among many of these breeds, within-breed genetic diversity is being undermined by the use of a few highly popular sires for breeding.

According to FAO's State of the World's Animal Genetic Resources report, at least one livestock breed a month has become extinct over the past seven years, which means its genetic characteristics have been lost forever.

Around 20 per cent of the world's breeds of cattle, goats, pigs, horses and poultry are currently at risk of extinction, according to the report, the first global assessment of livestock biodiversity and of the capacity of countries to manage their animal genetic resources.

"In this situation, the world cannot simply take a business-as-usual, wait-and-see attitude. Climate change means that we are entering a period of unprecedented uncertainty and crisis, which will affect every country," Mr. Müller said.

"Although animal genetic resources are important for everyone, they are particularly important for many livelihoods in developing countries, often of the very poorest," he added, stressing the need for governments to assist poor livestock keepers, who are the custodians of a large proportion of animal genetic diversity.

He highlighted climate change as a significant factor to be added to many other threats to livestock breeds. These include rapid, poorly regulated economic and social changes; increasing reliance on a small number of high-output breeds; animal diseases; and poverty, socio-economic instability and armed conflict in some of the areas richest in animal genetic resources.

Representatives of over 120 countries, including policy-makers, scientists, breeders and livestock keepers, are taking part in the week-long meeting to negotiate and adopt a Global Plan of Action for Animal Genetic Resources. The plan will comprise strategic priority areas as well as provisions for implementation and financing.

The last 10 years has seen an explosion of genetic studies aimed at improving the understanding of the origins and genetic characteristics of livestock genetic resources. Although the focus of most of this work has been Europe, remarkable progress has also been made in the developing world, thanks to the support of development agencies. Since 1995, ILRI has developed, in collaboration with National Agricultural Research Systems in developing countries, a dynamic programme on the molecular genetic characterisation of indigenous livestock of Africa and Asia. Molecular diversity information on cattle, sheep, goat, chicken, yak and OldWorld Camelidae are now partly available at country, regional and/or continental levels. The results show that the origin of the present day livestock diversity is more complex than previously thought with evidences for multiple origins or domestications. Livestock also show different patterns of geographic distribution of diversity in relation to the history of the domesticated species. These findings have direct applications to the design of strategies aiming to conserve diversity to maximise future utilisation. We have now the tools to understand the diversity of the genetic make-up of a livestock breed or population with direct applications to on-farm and on-station breeding programmes. Genetic characterisation may also provide a genetic signature for breed uniqueness within the context of intellectual property rights. Examples of applications of molecular diversity studies in livestock conservation and utilisation are illustrated by results obtained from the ILRI-led research.

Over the past 10,000 years, humans have domesticated about 40 species and created many thousands of breeds. For its survey, the agricultural organisation focussed on the 28 most significant of them, including horses, donkeys, cattle, camels, sheep, goats, swine, chickens, geese, Japanese quail, ducks and turkeys.

In addition to serving human needs, experts say those animals represent an important reservoir of genetic diversity, play a significant role in shaping a number of ecosystems, and are central to the history and culture of many communities.

Yet more than 600 breeds of livestock have already vanished, the organisation estimates, and another 78 are lost each year. Experts attribute the losses to the growing industrialisation of agriculture to meet rising demand around the world, particularly for poultry, pork and dairy cows; replacement of draft animals with machines; "indiscriminate" crossbreeding in an effort to improve indigenous stock; a narrow focus on certain breeds, like the Holstein for milk, to the exclusion of others; war and natural disasters; loss of family farms; and changing tastes in food and fashion.

Thus, for example, after the collapse of the market for mohair in the 1990's, the population of Angora goats, which produce it, plummeted in this country as farmers turned to other breeds.

In the 20th century, selective scientific breeding intensified and produced animals that were vastly better in specialised areas than others. Industrial hens bred purely to lay eggs will produce as many as 300 a year, experts say, compared with 20 to 30 for most indigenous birds. Industrial broiler chickens bred for meat will mature in 6 to 7 weeks, as opposed to 12 weeks for traditional stock. And the pork industry has developed pigs that mature earlier and produce leaner meat than their unrefined cousins.

12

Farm Animal Genomics

INTRODUCTION

Farm animal genomics is of interest to a wide audience of researchers because of the utility derived from understanding how genomics and proteomics function in various organisms. Applications such as xenotransplantation, increased livestock productivity, bioengineering new materials, products and even fabrics are several reasons for thriving farm animal genome activity. Currently mined in rapidly growing data warehouses, completed genomes of chicken, fish and cows are available but are largely stored in decentralised data repositories. In this paper, we provide an informatics primer on farm animal bioinformatics and genome project resources which drive attention to the most recent advances in the field.

Genomics is the scientific study of structure, function and interrelationships of both individual genes and the genome in its entirety. The field has evolved from identifying short nucleotide strings of DNA to the sequencing of an organism's complete genome. Current progress in genomics research has facilitated comprehensive mapping of the building blocks of biology. Ultimately, researchers hope to gain mastery over the fundamental description of cellular function at the DNA level. This would encompass gene regulation, in which proteins often regulate their own production or that of other proteins in a complex web of interactions. Databases can be developed to provide solutions to problems that people encounter when dealing with massive amounts of data.

Proteomics, or the science of protein structure and function, is now a hot spot in biomedicine. A key component to the next revolution in the 'post-genomic' era will be the increasingly widespread use of protein structure in rational experimental design. New computational methodologies now yield structure models that are, in many cases, quantitatively comparable to crystal structures, at a fraction of the cost. The technical challenge is the complete coverage of physico-chemical properties for thousands of proteins. Thus, by analytically investigating genes and proteins, researchers have developed the umbrella study of bioinformatics, the science of analysing biological data using cutting-edge computing techniques.

Bioinformatics deals with methods for storing, retrieving and analysing biological data, such as nucleic acid (DNA/RNA) and protein sequences, structures, functions, pathways and genetic interactions: the computational management of all kinds of biological information. Rather than merely a mixture of computer science, data management and genome sciences, Bioinformatics now encompasses both conceptual and practical tools for the propagation, generation, processing and understanding of scientific ideas and biological information.

DNA AND PROTEIN SEQUENCES DATA BANKS

The availability of many complete genome sequences from different species can bring insight into the function of conserved non-coding regions of DNA sequence. Organisation of the data into coding (genes) and non-coding sequences, in addition to organising these data into databases for 'DNA databases' and proteins (*See Table 12.1 on next page*), is central for bioinformatics. Parsing the sequence of DNA or protein is a first step, followed by curation then analysis and sometimes re-curation based on the analysis. The presence of these databases and derivative search engines gave rise to programmes, such as FASTA and PSI-BLAST, which are DNA and protein sequence analysis tools, respectively. These tools facilitate searches for sequences that resemble one another and homologous relationship inferences. Utilisation of similarity search programmes, which operate on sequences, has helped in augmenting annotated databases, which house information about sequence domains. Information recording, retrieval and cataloging continue to advance further database capabilities. Collaborating research centers, particularly in the US, Japan and the UK, have been actively collecting sequence data and making it accessible to public (Table 12.1).

Table 12.1: General genomics and proteomics databases: comprised of resources for human, goat, mouse, deer, rat and horse genomes

DB Name	DB type	Major contents
General utility databases		
Gen Bank	Sequence DB (for all organisms)	DNA/protein
PDB	The Protein Data Bank proteins structure	Experimentally determined 3D structure of proteins
EMBL	The EMBL nucleotide sequence database.	Nucleic sequences Nucleotide sequences of loci
Codon usage DB	Codon usage frequency	Codon usage in animals and other organisms
OMIA	Online mendelian inheritance in animals	Mendelian inheritance in animals
Entrez gene	Curated sequence and descriptive information about genetic loci. Genes–human	Loci sequences
OMIM	Online mendelian inheritance in man Genes–human	Catalog of human genes
GOBASE	The organelle genome database. Nucleic sequences mitochondrial	Mitochondrial genes
IMGT	The international immunogenetics database. Genes of multi-species	Genes evolving in immunology

(Contd...)

Swiss-Prot	Annotated protein sequence database. Proteins–sequences	Annotated sequences of proteins
Special genomic region databases		
ISIS	Introns of genes. Genes of multi-species sequences introns	Introns of genes
Loci-specific databases		
Deerbase	The deer genome database. Genomic mapping	Loci homologies with deer
Goatmap	Mapping the goat genome. Genomic mapping	Loci homologies with goat
MGD	The mouse genome database. Genomic mapping	Loci homologies with the mouse
Ratmap	The rat genome DB. Genomic mapping	Loci homologies with rat
Comparative genomic databases		
Homology form	The mammalian comparative map	Comparative maps
Homolo Gene	The homologene database	Gene homology

FARM ANIMALS: AN UNEXPLOITED GOLD MINE FOR BIOTECH

Farm animals are quite valuable as resources, often notable as models for pathology and physiology studies. Magnussen discusses how a variety of these farm animal models are used. The reproductive physiology of farm animals is more similar to humans than that of rodents because farm animals have longer gestation and pre-pubertal periods than mice. Specific farm animal physiology, such as the digestive system of the pig is similar to that of humans. These attributes of farm animals reveal that they are an unparalleled resource for research replicating human physiological function.

For decades, breeders have altered the genomes of farm animals by first searching for desired phenotypic traits and then selecting for superior animals to continue their lineage into the next generation. This genomic work has already facilitated a reduction in genetic disorders in farm animals, as many disease carriers are removed from breeding populations by purifying selection. By studying diverse phenotypes over time, researchers can now monitor mutations that occur as wild species become domesticated. Farm animal food safety will remain a concern for some time; however, advancements such as the discovery of Escherichia coli resistant genes in the pig can mediate most of the problems. Moreover, resources devoted to investigating the genomes of farm animals can bring eventual economic benefits. For example, isolation of DNA from animal tissue can be used as an inexpensive method for tracking the origin of a meat sample, providing the recipient with the quality assurance of that food.

There has been significant interest in the first complete analysis of the draft genome sequence of the chicken. This sequence has given rise to chicken genome array chips and a number of web based mapping tools. The great importance of the chicken as a scientific resource can be seen from the research on avian leucosis virus, which has led to developments in the areas of proviral insertion-mediated oncogenesis and vertebrate viral-host interactions. Still, many chicken lines are being lost due to facility downsizing and closings.

The release of the first draft of the chicken genome in March 2004 spawned the current boom in chicken genomic research (Table 12.2). From an evolutionary standpoint, investigation of the chicken genome will provide significant information needed to understand the vertebrate genome evolution, since the chicken is between the mammal and fish on the evolutionary tree. Furthermore, the chicken remains significant as a food animal which comprises 41 per cent of the meat produced in the world and serves as a reliable model for the study of diseases and developmental biology. With this sequenced genome, chicken breeders will have a framework for investigating polymorphisms of informative quantitative traits to continue their directed evolution of these species. The chicken genome is also effective as a comparative genomic tool that sheds additional light on various aspects of our own genome. In addition, complementary DNA microarrays for the chicken have already been produced for the study of metabolic and other systemic processes.

Table 12.2: Chicken-based genomic resources

Database	Content
Poultry and avian research resources	A detailed stock and curator listing of available avian research stocks are provided
Chicken variation database	An integrated information system for storage, retrieval visualisation and analysis of chicken variation data
Chicken genome browser	The genome browser zooms and scrolls over chromosomes, showing the work of annotators worldwide
NCBI chicken genome resources	Provides information on chicken-related resources from NCBI and the chicken research community
Wellcome trust chicken genome browser	This site presents an annotation of the first draft chicken genome assembly
WUGSC chicken genome site	Possesses help information on the chicken genome
Chicken genome array	The GeneChip chicken genome array is a key research tool for the study of chicken genomics and chicken viral pathogens
ChickCmap	ChickCmap allows the alignment of the different available maps in chicken
ChickFPC	Search using a sequence name, gene name, locus or other landmark

(Contd...)

Chicken database	Online public database browser
AvianNet	A portal to information on the chicken genome and chicken biology
ChickEst	Provides access to 339 314 Gallus gallus ESTs
US poultry genome project	Supported national animal genome research programme to serve the poultry genome mapping community
ChickAce	Mainly intended to store mapping information in chicken
University of Deleware ChickEST database	Contains over 40000 EST sequences from the chicken cDNA libraries in the UD collection
TIGR G.gallus gene index	Integrates research data from international G.gallus EST sequencing and gene research projects
Bacterial artificial chromosome (BAC)-based physical and genetic map	A genome-wide, BAC-based integrated genetic and physical map of the chicken genome
Consensus linkage map of chicken genome	Wageningen University chicken consensus linkage map
Chick RH server	RH mapping on INRA chicken radiation hybrid panel
Chicken-IMAGE	Improvement of chicken immunity resistance to disease based upon analysis of genome
Chickmap	Aim to construct an integrated genetic and physical map of the chicken genome
Chicken genome array	Key research tool for the study of chicken genomics and chicken viral pathogens
Chicken genome browser	Ensembl home ensembl chicken exportview
Chicken chromosome linkage map	Provides comparative mapping information

(Contd...)

Chicken BAC library	A BAC library of the chicken genome has been constructed
GEISHA	G.gallus (chicken) EST and in situ hybridization analysis database
NCBI chicken genome map viewer	G.gallus (chicken) genome

THE SWINE SEQUENCE

The sequencing of the pig genome generated an invaluable resource for advancements in enzymology, reproduction, endocrinology, nutrition and biochemistry research. Since pigs are evolutionarily distinct to both humans and rodents, but have co-evolved with these species, the diversity of selected phenotypes make the pig a useful model for the study of genetic and environmental interactions with polygenic traits. The sequencing of the pig genome is also instrumental in the improvement of human health. Clinical studies in areas such as infectious disease, organ transplantation, physiology, metabolic disease, pharmacology, obesity and cardiovascular disease have used pig models (*See Table 12.3 on next page*). In the near future, the sequencing of the porcine genome will allow gene markers for specific diseases to be identified, assisting breeders in generating pig stocks resistant to infectious diseases. Furthermore, as researchers investigate the swine genome and isolate genes that may impact the economics of breeding, members of the commercial pig industry are able to use this information to garner benefits.

THE BOS TAURUS GENOME

The mammalian order Cetartiodactyla (possessing B.taurus or cattle) is of great interest since it represents a group of eutherian mammals phylogenetically distant from primates. Working with the cow species, B.taurus, is significant because the cow is such an economically important animal. This form of livestock makes up the beef and milk production industry, which is one of the largest industries in the United States. The identification of numerous single-nucleotide polymorphisms (SNPs) makes it possible for geneticists to find associations between certain genes and cow traits that will eventually lead to the production of superior-quality beef. After the completion of the September 2004 B.taurus draft assembly, this genome has functioned as a vehicle for studies on non-primate and non-rodent genomes as well as in comparative genomics. Similar to the pig, the cow also serves as a good animal model for obesity, infectious diseases and female health.

Table 12.3: Genomic and proteomic study of the swine genome

Database	Content
NCBI pig genome resources	Brings together information on porcine-related resources from NCBI
Pig EST database	Pig EST database accommodates 98 988 pig ESTs, which were obtained from various sources
Genomic targets for comparative sequencing	Genomic targets of the pig
Pig QTL database	The database makes it possible to compare on pig chromosomes the most feasible location for a gene responsible for quantitative trait important to pig production
Swine genome maps	Swine gene mapping information
Porcine genome physical mapping project	A physical map of the porcine genome has been generated by an international collaboration of four laboratories
Cytogenetic map of the pig	List of genes mapped on pig cytogenetic map
PigBase	PigBase is a computer database that includes information on papers published about gene mapping in the pig
Pig genome mapping	Listing of pig genome databases
Blast pig genome	Blast pig sequences
Pig genome sequencing project	Procine genome project
Monsanto swine genome project	Produce genetic information from cDNA libraries made from swine tissue
ARK-genomics	Collaborative center for functional genomics in farm animals

SHEEP (OVIS ARIES) GENOME PROJECT

Bioinformatics researchers from New Zealand, US, UK and Australia have come together to work on the sheep genome map. The focal point of interest in sheep is based on the quest to maximize sheep meat and cotton wool production. This sector of the corporate farming industry is so intent on this biotechnology project that AWI, Meat and Livestock Australia and nine other partners are investing $50 million into the Sheep Genome Project to ensure its completion. As a resource in biological science, researchers have mapped a subset of genes that have also been mapped in humans and mice. These studies have revealed the existence of mutations that yield phenotypes unique to the sheep, demonstrating that genetic analysis of the sheep can enhance our knowledge of biological pathways in other mammalian species.

THE FISH GENOME PROJECT

Fish are one of the most studied organisms. Researchers have investigated the genome of zebrafish, medaka, pufferfish (Fugu rubripes and Tetraodon nigroviridis) and sticklebacks. The zebrafish genome has attracted the attention of various pharmaceutical and biotech companies owing to the ease with which scientists can use this fish to study the gene function. The number of databases and informatics resources related to fish genomic study on the Internet is consistently growing. Japanese researchers are interested in medakas from an evolutionary standpoint. Other fish genome sequences, such as the pufferfish genome are similar to the Fugu genome in size and are ~7.5 times smaller than the human genome. This has led many to believe that the genes conserved between these two species would reveal the minimal number of genes required for a vertebrate organism. Further investigation of the Fugu genome supported this hypothesis and showed the ability of the Fugu genome to aid in the study of vertebrate functional non-coding sequences. Other studies of fish have focussed on the stickleback, for its variable body shapes, ecology or its behaviour. In general, fish genomic work is of interest to the commercial fish farming community. Based on transgenic studies, antifreeze protein and fish growth hormone have been introduced into fish genomes, creating fish with greater cold tolerance and faster growth rates.

THE FORGOTTEN RABBIT

Little genomics attention is paid to rabbits in comparison with other animal models, such as the mouse, rat and fruit fly in the pre- and post-genomic era. However, experimental models, such as the Alba or the 'mighty lighting rabbit', were developed as glowing mutants that shine under special light for commercial reasons. French scientists created Alba using a process called zygote microinjection. In this process, the scientists plucked a fluorescent protein from the fluorescent jellyfish Aequorea victoria. Then, they modified the gene to make its glowing properties twice as powerful. This gene, called EGFG (enhanced green fluorescent gene, was then inserted into a fertilised rabbit egg cell that eventually grew into Alba (Amanda onion). Debates about the project itself and about the practice of manipulating genes in animals for research have quickly arisen in the research community. The French National Institute of Agronomic Research hesitated to release these rabbits owing to protests over its development.

Chinese scientists have placed rabbit genes in cotton plants, producing cotton fibers as bright and soft as rabbit hair but stronger and warmer. This indicates not only the rise of the rabbit in experimental models for genetics engineering, but also for future medical experimentation models. Despite the controversy that the 'shining rabbit' and the 'cotton rabbit' raised, it is clear that this type of genetic engineering is at an early stage.

FARM ANIMAL GENOMICS: CURRENT STATISTICS

As researchers delve into the composition of farm animal genome sequences, new functional and biological data emerge.

One of the major sources of information on farm animal genomics is the ArkDB, which is available through the Roslin Institute (UK) and Texas A&M University (USA). The ArkDB provides detailed genomic mapping data on sheep, chicken, cow and pig genomes, including data on PCR primers, genetic linkage map assignments of specific loci and markers, and cytogenetic map assignments. Quantitative functional information, such as the number of clones, microsatellites and associated mapping assignments, can lend insight into the complexity of the models available. For example, the number of primers tabulated in the swine genome currently far exceeds that known for the cow, indicating more varieties of genes are available for study in the swine. Much of this information remains at an early stage and with increasing experimentation, compilation and analysis, will be refined.

QUANTITATIVE TRAIT LOCI AND GENETIC LINKAGE

One of the primary challenges in modern biology is the understanding of the genetic basis of phenomic diversity within and among species. The foundation for this diversity lies in genetic governance of both how traits are expressed and the associated linkage maps. Genes mapped in some farm animals serve as 'anchors' across the comparative maps of other species. Quantitative trait loci (QTLs) play a major role in farm animals and the related biotechnology industry, as they can further the identification of traits related to meat and milk production. A number of studies have been conducted to detect QTLs that can be used for determining gene variances. The third generation of sheep linkage map contains 1062 loci (941 anonymous loci and 121 genes) and is a compilation of genotype data generated by 15 laboratories using the IMF population. The ArkDB provides current summaries of linkage and cytogenetic map assignments, polymorphic marker details, PCR primers and two point linkage data. The ArkDB also is a major source for mapped sheep loci, with SheepBase containing almost 1500 loci. The Roslin Institute has developed the 'resSpecies' database to study genetic linkage maps, QTLs, alleles and other markers. Bovine QTL databases are available from several sources. Various traits, such as milk yield and composition, are available along with QTL dot maps, which provide varied positional and statistical information. The pig QTL includes two release versions from NCBI in May 2004 and from NAGRP in December 2004. The pig QTL database, or pig QTLdb, contains all published pig QTL data from the past decade. The user can locate genes responsible for quantitative traits central to pig production. To date, 791 QTLs from 73 publications, representing 219 different traits are incorporated into the database in the first release and 1129 QTLs, from 86 publications representing 235 different traits in the second release.

GENOME POLICIES FOR TRANSGENIC ANIMALS

The federal government's role as a legal authority for protecting the environment from scientifically introduced transgenic organisms has not been clearly outlined. This underscores the need for greater governmental attention to address legal problems that will arise as animal biotechnology continues to grow. Many argue that it is not only a moral imperative to respect the intrinsic sentience of animals but also a legal one. The European Union's Treaty of Amsterdam states that animals must be treated humanely, as sentient beings, which some argue implies respect for their intrinsic nature and protection against such infringements of that respect in studies such as transgenic animal experiments . Furthermore, US food regulatory agencies, such as the United States Department of Agriculture and the Food Safety and Inspection Service, charged with the duty of making sure that transgenic animals anticipated for human consumption are correctly labeled and wholesome, have already moved forward to institute policies regulating the slaughtering of non-transgenic animals created from transgenic animal experiments.

ETHICAL RESERVATIONS OF FARM ANIMAL GENOMIC STUDY

As farm animal species continue to be sequenced (one of the latest being the cow, B.taurus), farming companies are using this information to perform genetic profiles of traditionally bred animals and to genetically engineer or clone other animals. Although many farmers may enjoy access to greater genetic resources, animal rights organisations oppose the use of animals such as transgenic goats that produce silk proteins used to make Biosteel fibers because of the allegedly inhumane treatment of animals during these studies. Proponents of such experiments point to the ability of transgenic mouse mammary glands to assemble and secrete recombinant human fibrinogen. These studies raise many concerns about research ethics and species integrity.

THE POTENTIAL OF FARM ANIMAL GENOMICS

Farm animal genomic studies continue to attract audiences excited by the multitude of applications. The meat industry can now use cow and chicken genomic data to confirm the quality of meat products. For example, meat producers can now confirm the parentage of an Angus cattle breed by performing a genetic blood test or attempt to identify the SNPs associated with high-quality beef. Other companies are using genomic information to determine disease-resistant genes in shrimp and then are selectively mating the shrimp that carry them in order to create disease resistant strains. In the healthcare arena, farm animal genomic work will aid in enterprises such as xenotransplantation (the transfer of animal tissues or organs into humans). Though animal organs may be used someday to satiate organ donor shortage, genomic work in this area is still in its early stage. Many of the immediate practical applications of farm animal genomics show potential for growth in this field.

13
Recombinant DNA Technique

INTRODUCTION

Recombinant DNA is a form of artificial DNA that is engineered through the combination or insertion of one or more DNA strands, thereby combining DNA sequences that would not normally occur together. In terms of genetic modification, recombinant DNA is produced through the addition of relevant DNA into an existing organismal genome, such as the plasmid of bacteria, to code for or alter different traits for a specific purpose, such as immunity. It differs from genetic recombination, in that it does not occur through processes within the cell or ribosome, but is exclusively engineered.

The Recombinant DNA technique was engineered by Stanley Norman Cohen and Herbert Boyer in 1973. They published their findings in a 1974 paper entitled "Construction of Biologically Functional Bacterial Plasmids in vitro", which described a technique to isolate and amplify genes or DNA segments and insert them into another cell with precision, creating a transgenic bacterium. Recombinant DNA technology was made possible by the discovery of restriction endonucleases by Werner Arber, Daniel Nathans, and Hamilton Smith.

Because of the importance of DNA in the replication of new structures and characteristics of living organisms, it has widespread importance in recapitulating via viral or non-viral vectors, both desirable and undesirable characteristics of a species to achieve characteristic change or to counteract effects caused by genetic or imposed disorders that have effects upon cellular or organismal processes. Through the use of recombinant DNA, genes that are identified as important can be amplified and isolated for use in other species or applications, where there may be some form of genetic illness or discrepancy, and provides a different approach to complex biological problem solving.

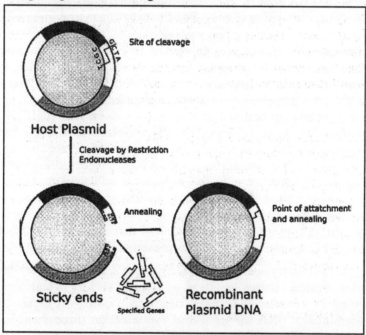

Fig. 13.1: A simple example of how a desired gene is inserted into a plasmid. In this example, the gene specified in the white colour becomes useless as the new gene is added

CLONING AND RELATION TO PLASMIDS

The use of cloning is interrelated with Recombinant DNA in classical biology, as the term "clone" refers to a cell or organism derived from a parental organism, with modern biology referring to the term as a collection of cells derived from the same cell that remain identical. In the classical instance, the use of recombinant DNA provides the initial cell from which the host organism is then expected to recapitulate when it undergoes further cell division, with bacteria remaining a prime example due to the use of viral vectors in medicine that contain recombinant DNA inserted into a structure known as a plasmid.

Plasmids are extrachromosomal self-replicating circular forms of DNA present in most bacteria, such as Escherichia coli (E. Coli), containing genes related to catabolism and metabolic activity, and allowing the carrier bacterium to survive and reproduce in conditions present within other species and environments. These genes represent characteristics of resistance to bacteriophages and antibiotics and some heavy metals, but can also be fairly easily removed or separated from the plasmid by restriction endonucleases, which regularly produce "sticky ends" and allow the attachment of a selected segment of DNA, which codes for more "reparative" substances, such as peptide hormone medications including insulin, growth hormone, and oxytocin. In the introduction of useful genes into the plasmid, the bacteria are then used as a viral vector, which are encouraged to reproduce so as to recapitulate the altered DNA within other cells it infects, and increase the amount of cells with the recombinant DNA present within them.

The use of plasmids is also key within gene therapy, where their related viruses are used as cloning vectors or carriers, which are means of transporting and passing on genes in recombinant DNA through viral reproduction throughout an organism. Plasmids contain three common features — a replicator, selectable marker and a cloning site. The replicator or "ori" refers to the origin of replication with regard to location and bacteria where replication begins. The marker refers to a gene that usually contains resistance to an antibiotic, but may also refer to a gene that is attached alongside the desired one, such as that which confers luminescence to allow identification of successfully recombined DNA. The cloning site is a sequence of nucleotides representing one or more positions where cleavage by restriction endonucleases occurs. Most eukaryotes do not maintain canonical plasmids; yeast is a notable exception. In addition, the Ti plasmid of the bacterium Agrobacterium tumefaciens can be used to integrate foreign DNA into the genomes of many plants. Other methods of introducing or creating recombinant DNA in eukaryotes include homologous recombination and transfection with modified viruses.

CHIMERIC PLASMIDS

When recombinant DNA is then further altered or changed to host additional strands of DNA, the molecule formed is referred to as "chimeric" DNA molecule, with reference to the mythological chimera, which consisted as a composite of several animals. The presence of chimeric plasmid molecules is somewhat regular in occurrence, as, throughout the lifetime of an organism, the propagation by vectors ensures the presence of hundreds of thousands of organismal and bacterial cells that all contain copies of the original chimeric DNA.

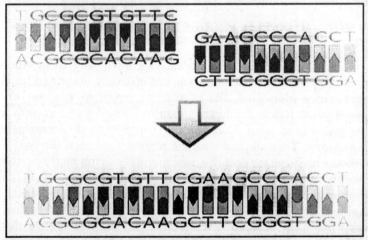

Fig. 13.2: **An example of chimeric plasmid formation from two "blunt ends" via the enzyme, T4 Ligase**

In the production of chimeric plasmids, the processes involved can be somewhat uncertain, as the intended outcome of the addition of foreign DNA may not always be achieved and may result in the formation of unusable plasmids. Initially, the plasmid structure is linearised to allow the addition by bonding of complementary foreign DNA strands to single-stranded "overhangs" or "sticky ends" present at the ends of the DNA molecule from staggered, or "S-shaped" cleavages produced by restriction endonucleases.

A common vector used for the donation of plasmids originally was the bacterium Escherichia coli and, later, the EcoRI derivative, which was used for its versatility with addition of new DNA by "relaxed" replication when inhibited by chloramphenicol and spectinomycin, later being replaced by the pBR322 plasmid. In the case of EcoRI, the plasmid can anneal with the presence of foreign DNA via the route of sticky-end ligation, or with "blunt ends" via blunt-end ligation, in the presence of the phage T4 ligase, which forms covalent links between 3-carbon OH and 5-carbon PO4 groups present on blunt ends. Both sticky-end, or overhang ligation and blunt-end ligation can occur between foreign DNA segments, and cleaved ends of the original plasmid depending upon the restriction endonuclease used for cleavage.

SYNTHETIC INSULIN PRODUCTION USING RECOMBINANT DNA

Until the 1920s, there was no known way to produce insulin because the hormone was not officially identified until 1921. Once identified, the production problem was quickly solved when it was found that insulin from the pancreas of a cow, pig or even some species of fish could be used successfully in humans. This method was the primary solution for type 1 diabetes mellitus for decades, and manufacturing methods had steadily improved the purity of the hormone which was made from animal pancreases. However, proponents of the genetic engineering technology continued to raise what they claimed was a looming problem with traditional methods of insulin production: a supposed shortage of supply in the not-too-distant future. But in the 1987 book "Invisible Frontiers: The Race to Synthesize a Human Gene", author Stephen S. Hall wrote that the supposed shortage is now known to be an assumption based on mistaken facts. He wrote:

To hear some tell it, there was never a supply problem with pig pancreases in the first place. "The whole thing was rubbish," insists Paul Haycook, research director at Squibb-Novo. "There was never a shortage of pig pancreases, and there never will be." Haycook blames the scare on a miscalculation by an official who had prepared projections for the Food and Drug Administration — a mistake based, ironically, on a mistake in an Eli Lilly training brochure which confused kilograms with pounds. Instead of projecting an insulin shortage by 1982, a revised FDA report predicted adequate insulin supplies through the year 2006. In any event, there is never likely to be a shortage caused by a scarcity of pancreases.

Scientists and entrepreneurs were very eager to prove they could devise another way to synthesise the hormone, in part, because of competition from other researchers and also because of the promise for the fame and fortune that its so-called "discovery" could bring them. Insulin was part of a wider vision to introduce biotechnology medicines, and was chosen specifically because it is a simple hormone and was therefore relatively easy to copy. However, the motive was never to improve the lives of people with diabetes, but to prove that the technology worked. Insulin was chosen as the ideal candidate because it is a relatively simple protein, it was so widely used that if researchers could prove that biosynthetic "human" insulin was safe and effective, then the technology would be accepted as such, and it would open the flood gates for many other products to be made this way, along with millions of dollars.

That was exactly what happened. One of the biggest breakthroughs in recombinant DNA technology happened in the manufacture of biosynthetic "human" insulin, which was the first medicine made via recombinant DNA technology ever to be approved by the FDA.

Henry I. Miller was an early advocate of biotechnology and drugs, and continues to be so as a fellow at the Hoover Institute even though he no longer works for the FDA. Miller began work for the FDA as head of a special department created to establish new procedures for approving drugs created through biotechnology. In his book "To America's Health:

A Model for Reform of the Food and Drug Administration" (Hoover Institution Press, 2000), Miller states that he pushed for rapid approval of biosynthetic insulin from his boss, who was not comfortable approving it on such short notice, especially when it had been tested on so few people. Amazingly, Miller admits that he actually waited for his boss to go on vacation, and then took the approval to his boss' boss, who then approved the drug.

As far as technical details, the specific gene sequence, or oligonucleotide, that codes for insulin production in humans was introduced to a sample colony of E. coli (the bacteria found in feces). Only about 1 out of 106 bacteria picks up the sequence. However, this is not really a problem, because the lifecycle is only about 30 minutes for E. coli. This means that in a 24-hour period, there may be billions of E. coli that are coded with the DNA sequences needed to induce insulin production.

However, a sampling of initial reaction showed that Humulin was greeted more as a technological rather than a medical breakthrough, and that this sentiment was building even before the drug reached pharmacies. As early as 1980, the British magazine New Scientist reported, "Other big chemical manufacturers predict that Eli Lilly's massive $40 million investment in two plants to make insulin - may be a classic example of backing a loser."[citation needed]

The Economist concluded: "The first bug-built drug for human use may turn out to be a commercial flop. But the way has now been cleared-and remarkably quickly, too—for biotechnologists with interesting new products to clear the regulatory hurdles and run away with the prizes."[citation needed]

Ultimately, widespread consumer adoption of biosynthetic "human" insulin did not occur until the manufacturers removed highly-purified animal insulin from the market.

14

Anaerobic Digestion in Animals

INTRODUCTION

Anaerobic digestion consists of a series of reactions which are catalyzed by a mixed group of bacteria and through which organic matter is converted in a stepwise fashion to methane and carbon dioxide. Polymers such as cellulose, hemicellulose, pectin, and starch are hydrolysed to oligomers or monomers, which are then metabolised by fermentative bacteria with the production of hydrogen (H_2), carbon dioxide (CO_2), and volatile organic acids such as acetate, propionate, and butyrate. The volatile organic acids other than acetate are converted to methanogenic precursors (H_2, CO_2, and acetate) by the syntrophic acetogens. Finally, the methanogenic bacteria produce methane from acetate or from H_2 and CO_2. Stable digestor operation requires that these bacterial groups be in dynamic equilibrium.

Generally, the final products of microbial degradation of carbonaceous material in an anaerobic ecosystem are methane and carbon dioxide, which are both odourless. However, when wastes are stored, the rate of methane production is not fast enough to prevent the accumulation of products of the acid-forming fermentation. In other words, the acid-forming and methanogenic steps in the microbial degradation of stored organic matter are unbalanced. The imbalance between the processes of acid fermentation and methane production is the key to understanding the accumulation of volatile malodorous products. Under balanced conditions, the volatiles are converted to methane and carbon dioxide.

In many storage systems for livestock manures, therefore, an unbalanced fermentation is created and objectionable odours result from the accumulation of volatile malodorous intermediates. However, in an anaerobic digestion system designed and operated for methane production, the two phases of acid fermentation and methane production are kept in balance and many odorants are degraded, resulting in lower concentrations in the digestor effluent.

Although anaerobic decomposition is considered to be the source of manure odour, it also offers the potential for reducing odourants if they can be contained sufficiently during the decomposition process. Odour control is the primary goal of some anaerobic digestion systems installed on livestock operations. In addition, anaerobic digestion has other benefits such as waste stabilisation and iquefaction, and production of biogas, an alternative fuel for on-farm (or off-farm) use. The fertiliser value of the raw manure is conserved in the digested effluent and there is also a considerable reduction in the number of pathogens.

ODOUR IMPACT

Anaerobic digestion has been shown to reduce the offensiveness of manure odour Welsh et al. demonstrated that anaerobic digestion reduced the presence and offensiveness of swine manure odours. Also, anaerobic digestion at 35°C was more effective in controlling odor than at 25°C provided that solids retention times were 10 days or greater. The reduced presence and offensiveness of digested-effluent odours remained even after storage for nearly three months. Digested pig slurry, when spread on land, produced an odour significantly less than that from undigested slurry. During the first six hours after spreading, when the rate of odour emission is highest, digestion reduced emission by 79 per cent. Also, additional storage did not negate the benefits of anaerobic digestion in reducing odour offensiveness from pig slurry applied to grassland.

Powers et al. have shown that effluent from anaerobic digestion of simulated flushed dairy manure was much less odourous than the fresh manure influent. Maintaining longer hydraulic retention times and separating the most fibrous solids by screening the manure prior to digestion further reduced odours, as evaluated by a human panel.

Anaerobic digestion was developed originally because of its ability to control and eliminate the malodour associated with domestic sludges . With environmental pressures increasing, there is a clear opportunity to implement the process for odour control of livestock wastes. Currently, only aerobic treatment offers similar benefits. However, the operational costs and complexity of aerobic treatment systems are greater than for anaerobic systems. Compared to conventional aerobic methods, which consume energy and produce large amounts of sludge requiring disposal, anaerobic treatment processes are net energy producers and produce significantly less sludge.

DIGESTOR DESIGN

A digestor design must be selected on a site-specific basis to match the waste management system of the individual operation and meet the needs of the producer. Anaerobic lagoons are often used for animal waste systems. They serve as storage facilities and achieve considerable solids breakdown. Although not free of odours, they are seldom the cause of an odour problem. A properly designed and operated lagoon system, where the hydraulic retention time (HRT) exceeds 60 days, will not have malodor problems during the year, except possibly in spring when the lagoon temperature rises and bacterial action increases. Most odours associated with anaerobic lagoons are caused by overloading. Frequently, the number of animals and the volume of manure produced have increased since the original lagoons were built, without a corresponding expansion of lagoon capacities. Use of a gas-tight cover allows capture of biogas for use as an energy source, although biogas production tends to vary seasonally due to temperature fluctuations. Moreover, any anaerobic lagoon (covered or not) is impractical in regions with a high water-table because of the potential for groundwater contamination.

Being a completely closed system, an anaerobic digestor allows more complete digestion of the odourous organic intermediates found in stored manure to less offensive compounds. Also, from an aesthetic perspective, digestor-based systems are preferable to open lagoons. The two most popular types of on-farm digestors in the United States are continuously stirred tank reactors (CSTR) and plug-flow digestors. The CSTR allows for complete mixing of digestor contents.

Continuous or intermittent feeding of fresh waste, usually as a slurry, is accompanied by overflow of digested waste to a storage area. The requirement for mechanical mixing dictates use of waste with a solids content in the range of 5-14 per cent total solids (TS). Stable CSTR operation requires HRTs of 15-30 days. Plug-flow digestors are unmixed systems in which waste flows semicontinuously as a plug through a orizontal reactor. The reactor may be an in-ground tubular tank or a covered, concrete-lined trench. HRTs are relatively long (c. 30 days) and a comparable solidscontent waste can be used as compared to the CSTR.

Digestor designs which require long retention times for effective treatment are, however, unsuited to flushed-manure systems where large volumes of dilute waste result, due to the capital costs of constructing sufficiently large digestors. In CSTRs, the solids retention time (SRT) is the same as the HRT. Because of the slow growth rates of syntrophic and methanogenic bacteria, reduction of the HRT in CSTRs risks causing washout of the active biomass, with consequent process failure. This problem can, however, be overcome by maintaining the microbial population within the digestor independently of the waste flow, i.e. by devising a means to maintain long SRT even at low HRT.

Recent developments in digestor design have focussed on retention of the active microflora within the reactor. These designs rely on the tendency of the bacteria involved to adhere to inert surfaces, forming biofilms, or to aggregate in settleable flocs or granules, and provide improved process stability and control. The fixed-film anaerobic reactor is one such design.

Active biomass retention permits reduction of the HRT from the 20-30 days characteristic of conventional anaerobic reactors to periods ranging from several hours to several days. Reduction of the HRT also implies considerable initial capital cost savings due to the decreased size requirements for the reactor.

The fixed-film anaerobic reactor immobilises bacteria on a packing material within the reactor, thereby preventing washout of microbial biomass. The principle of operation is that wastewater is passed through a column filled with an inert packing material. The packing material, also called media, acts as a surface for the attachment of microorganisms and as an entrapment mechanism for unattached flocs of organisms. This attached and entrapped anaerobic biomass converts both soluble and particulate organic matter in the influent wastewater to methane and carbon dioxide as the flow passes through the column. The media is fully submerged and wastewater flow can be in either upflow or downflow mode.

APPLICATION

Case studies of operating anaerobic digestors, including project and maintenance histories, have recently been compiled. Although previous research has been primarily concerned with the energy aspects of anaerobic digestion, a digestor functions as an integral part of the total waste management system and its advantages and disadvantages should be reviewed in the context of the overall system. Odour control is a particularly important consideration. Increasing public concern over odourous emissions from animal husbandry is bringing about a reappraisal of the feasibility of anaerobic digestion for odor control.

Since the application of fixed-film digestion to treat livestock waste at full-scale is relatively new, there are no set design parameters. Therefore, a pilot-scale facility consisting of four 90-gallon, continuously fed fixed-film digestors was constructed at the site to establish guidelines for the design, start-up and operation of the full-scale digestor. The major areas of concern were evaluating candidate media for performance and clogging potential, gaining information on biogas production and composition, and obtaining start-up and performance data. The pilot-scale digestors also serve to provide valuable hands-on materials-handling experience with the dairy wastewater.

Two pilot-scale digestors were packed with modular cross-flow plastic media and two were packed with vertical-flow plastic pipe media. All four digestors were started up using unamended dairy wastewater inoculum and were operated at 3-day HRT, at ambient temperature. In order to compare the effect of flow direction on media performance, one digestor for each media type was fed upflow and the other in downflow mode.

Results indicate that both media have similar performance characteristics at pilot-scale. At 3-day HRT, COD removal was 57 and 77 per cent for total and soluble COD, respectively, with a COD conversion efficiency of 0.35 m_3CH_4/kg COD removed. Average biogas composition was 78 per cent methane, 22 per cent carbon dioxide, and about 2,500 ppm hydrogen sulfide. After one year of operation, neither media showed evidence of potential clogging problems due to accumulation of fibrous solids. Also, solids build-up at the base of the reactors did not appear to be a problem, regardless of media type or flow direction.

In order to quantify the amount of odour reduction a fixed-film digestor can achieve, a threshold odour test (APHA, 1995) was conducted on the undigested influent and the digested effluent of the pilot-scale digestors. The test consists of making a series of dilutions of the wastewater with odour-free water until the sample no longer has a detectable odour, as perceived by the members of a human panel. The ratio by which the sample has been diluted is called the "threshold odour number" (TON). The higher the TON, the more odorous is the sample.

CONCLUSION

As the incidence of odour-nuisance complaints increases, livestock and dairy producers must seek an effective means of controlling odourous emissions. Anaerobic digestion under controlled conditions offers producers a holistic solution that allows them to coexist with their neighbours without limiting the enterprise. The process can be adapted to an individual situation and incorporated into an already existing manure management scheme. Addition of an anaerobic digestor allows adequate digestion of odourous organic intermediates found in stored manure to less offensive compounds. Odour panelists have repeatedly determined that digestor effluent has a much improved odour quality compared to fresh or undigested manure. Additional benefits such as biogas production, conservation of fertiliser value, pathogen reduction, resource recovery, and confinement of volatile emissions of environmental concern, make installation of an anaerobic digestor an increasingly attractive alternative to current waste management practices at many livestock operations.

15
Animal Proteomics

INTRODUCTION

Proteomics comprises the study and characterisation of cellular regulation: that happens to all the proteins in a given cell type, tissue or organism. It includes their expression, post-translational modifications, regulation etc. These are complex and costly works undertaken for man, for a few model species (rat, mouse, some microorganisms) and for organisms with a clear commercial relevance. However, most of the organisms that are a part of the human diet have not been submitted to proteome analyses: there are for example thousand of different species of fish consumed by man in different parts of the world. Fortunately, man proteins have conserved amino acid sequences, pI and molecular mass, which permits to apply the results from the model species to other ones, specially if the model and the novel one are closely related.

Globalisation, opening of international markets, consumer demands and clear differences between quality/price in different markets, permit the intervention of opportunistic elements and the falsification of the documentation that must accompany foodstuffs, usually in order to increase the profit. The consequences can be very serious, including death, in case a person allergic to a certain component consumes a product that contains it but is wrongly labelled and the component is not mentioned. The consumer is entitled to information about the species and origin of the components and the degree of pre- and processing (whether it has been frozen, cooked, irradiated, high pressure treated, etc). These data are also relevant to examine the degree to which microorganisms, especially pathogens, and parasites may have been eliminated from the product. Currently, there are no analytical methods to ensure whether most of these data – origin, variety, type and degree of processing, etc – are correctly declared or not.

PROTEIN DATABASE OF DROSOPHILA

A general catalogue integrated by 1184 [35S] Methionine + [35S] Cysteine labelled polypeptides from wing imaginal disc has been obtained. The level of expression for all the proteins has been quantitatively determined. The quantitative reproducibility of the analysis system has been estimated and all the controls studied as database reference to interpret the results of experiments with mutant discs. This enables us to generate comparative parameters for the study of proliferation, morphogenesis and differentiation of Drosophila and opens the possibility of rapidly defining the nature and quantity of changes in pattern of gene expression in developmental genetic studies.

In this way have compare the patterns of protein synthesis of wing, haltere, leg 1, leg 2, leg 3 and eye antenna imaginal discs of the late third instar larvae. Two dimensional gel electrophoresis followed by microsequencing has been used to purify and identify one of them. The microsequence data showed identity with amino acids encoded by the human proliferation association gene, pag, which is a thiol-specific antioxidant. By virtue of this homology we have cloned and sequenced two cDNAs that seems to define the peroxiredoxin family of Drosophila.

The database has been also used to evaluate changes in the patterns of protein synthesis in wing imaginal discs from two Drosophila mutants with abnormal wing disc development: fat (ft) and two different alleles of lethal giant disc (lgd).

APPLICATION OF PROTEOMICS TO STUDIES ON PHYLOGENY AND EVOLUTION

The matter dealing on zoology and proteomics are 2-D gel electrophoresis related to genetic variability, evolution and taxonomy. While two dimensional protein electrophoresis was originally used for systematic purpose it is particulary useful for studying phylogenetics and detection of evolutionary changes. 2-D gel databases as bioinformatic tools are applied and their role is changing as zoologists move toward protein expression rather than position variability in two-dimensional maps. Phylogenetic analysis based on DNA sequence information has transformed the current view of organic evolution. Woese et al. reclassify the traditional five kingdom (Monera, Prototista, Fungi, Plantae and Animalia) into three domains and this proposal has been alternatively evaluated based on three approaches (gen order, gene content, hierarchical classification) which use entire genome for phylogenetic reconstruction. Gene order and gene content methods are based on genetic distance criteria while hierarchical classification of proteome, through ORFs conversion, is based on overall statistical similarity. However, comparison of proteomes could be evaluate according to the principles and methods of Phylogenetic Systematics (cladism) expressing three different criteria of resemblance: monophyly, paraphyly and polyphyly. To construct parsimony based phylogenetic trees, data obtained by application of proteomics might be treated by step matrices method as best criteria, if they are protein sequences, and maximum parsimony or character compatibility if data are in binary form (presence or absence proteins). Use of cladistic approach (ancestor descendants sucession) would allow detect and characterize (mass spectrometry, sequencing) specific proteins which are unique features (synapomorphies) with evolutionary values. The anagenetic changes in the same tree and how these changes manifest during taxa evolution would show the reconstruction of ancestral characters states for each taxa. Comparative proteome analysis permits large-scale screening for protein genetic differences and by means of mass spectrometry, direct identification of proteins in sequence databases of entire organisms. That is the case of the rhabditid nematode Caenorhabditis elegans. Soil and free-living nematodes are continuously submitted to environmental stress being nematode-bacteria differential interaction a common fact in soil. Bacterial infection can be determinant for nematodes evolution as can be inferred by adaptation to such selective pressure of some nematode strains. Epidemiological experiment confirms that this adaptation has been fixed while application of proteomics detects the proteins differences between strains.

ANTIGEN PRESENTATION STUDIES USING MASS SPECROMETRY TECHNIQUES

The function of MHC molecules is to bind antigenic peptides generated by the antigen presenting cell (APC) processing mechanism and to present them to the T lymphocytes. For these peptides being correctly processed and presented, the help of molecules such as the invariant chain (Ii) and HLA DM is needed. We have studied the influence of the molecules Ii and DM in the class II antigen presentation pathway. We compare the different antigenic peptides presented by rat insulinoma epithelial cells, RINm5F, which are transfected with different combinations of the molecules in study (DRB1*0401, Ii and HLA-DM). Also, we check if those peptides fit the MHC union motive. The identification of these peptides is carried out following the protocol indicated in the fig. 15.1. (*See on next page*)

With the results obtained from the analysis by MALDI-TOF and after the characterisation by nanoESI-MS/MS of approximately 130 peptide sequences belonging to a total of 59 different proteins, we postulate the possible existence of alternative pathways of antigenic processing and presentation to the actually described.

PROTEOMICS ON THE TOTIPOTENT PLANARIAN STEM CELL

The planarian (phylum Platyhelminthes; class Turbellaria; order Seriata; suborder Tricladida) as a model organism has some characteristics that makes it very interesting for the investigation of developmental biology,

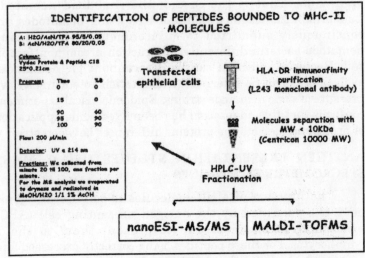

Fig. 15.1

regeneration, and the "up to date" investigation on stem cells. They are characterised by having totipotent stem cells, the neoblast cells, which are the only proliferative cell type present during the entire planarian life cycle. This gives the planarians a huge morphogenetic plasticity such as complete regeneration of any planarian piece, and continuosly shrink and growth of the planarian depending on food availability.

To date, proteomic studies carried out on planarians are very limited. The mentioned proteomic characterisation of the neoblast cells is at its first stage, which involves the dissociation of the planarian cells, separation of the cells previously labelled with calcein and fluorescein by flow cytometry and isolation of the neoblast cells by cell sorting. Future work wished to be accomplished consists in a characterisation of proteins responsable for the totipotent stage of the neoblast cells, for the proliferation capacity, and proteins responsable for the differentiation of the neoblasts into the different cell types found in the adult planarian. The characterisation includes the sequencing of the proteins followed up by functional studies of the genes corresponding to the proteins by RNA interference and transgenesis.

GENOMICS AND PROTEOMIC

Genomics and proteomics have largely enhanced our knowledge of the molecular basis which determine the different roles played by complex organisms. Viruses have developed strategies to alter key roles of infected cells and host responses that allow replication, assembling and dissemination. Therefore, their genetic material must necessarily contain information aimed to interfere in specific intracellular signalling pathways.

The full sequence of African Swine Fever Virus (ASFV) has been already described, but to date little is known about the function of most proteins codified by its genome, and a comprehensive understanding of its interaction with target cells is not yet at reach. This study will clear up the fundamentals of ASFV biology.

2D-E analysis of Vero-infected cells revealed unambiguous changes in cellular protein patterns since early times after infection. These changes were consistently found during infection with several viral strains. Modified proteins were further identified by their Peptide Mass Fingerprint (PMF) using the MASCOT platform and the NCBInr database. Major changes were produced in redox-related proteins, nucleoside diphosphate kinases, members of the Ran-Gppnhp-Ranbd1 complex or apolipoproteins. These cellular protein modifications have been described in other viral models and could represent distinct roles during infection related with apoptosis and transcriptional modulation mechanisms. The results obtained by this analysis in ASFV-infected cells will be discussed.

CELL PROLIFERATION

Hepatocytes, the major cell type in liver, are quiescent cells, but they retain the ability to proliferate in response to acute hepatic injury. After partial hepatectomy (PH) in which 66 per cent of liver mass was removed, all cells enter into the cell cycle in more or less synchronised manner. This is a good method to study cell cycle in an in-vivo model.

Cell proliferation is regulated by a family of serine-threonine kinases named cyclin-dependent kinases (CDKs).

CDK activity is regulated by a variety of mechanisms, including association with regulatory subunits named cyclins, phosphorylation of positive and negative regulatory sites, and binding to a number of proteins called CDK inhibitors (CKIs). The cell cycle inhibitor p21CIP1 is a CKI that regulates liver regeneration by modulating the activity of CDKs. Recently different proteins binding to p21CIP1 have been reported in different complexes, indicating that p21CIP1 could play different roles in different signal pathways.

To study the role of p21CIP1 during cell cycle, a proteomic approach of nuclear hepatocyctes extracts from knockout p21CIP1 compared to normal mice has been done. Hepatocyctes from liver mice where synchronised by PH and analysed at 0h. as control of quiescent cells and at 48h. after PH as a maximum DNA synthesis.

Here we report that knockout p21 -/- mice have different expression of nuclear proteins during cell cycle. Few of them have been characterised by MALDI-TOF after high resolution bidimensional gel electrophoresis separation of nuclear extracts of hepatocytes.

APPLICATIONS ON PROTEOMICS TO ANIMAL PHYSIOLOGY

Proteomics is a collection of scientific approaches and technologies aimed at characterising the protein content of cells, tissues, and whole organisms and one of its goals is to understand the biological role of specific proteins. The utilisation of these forward-thinking approaches will provide new insight into animal function. We present a review with some examples of actual and potential applications of proteomics in different fields of animal physiology paying special attention to environmental physiology of invertebrates.

16
Animal Cell Biotechnology

INTRODUCTION

Animal cell biotechnology constitute an important field in Biotechnology nowadays. Animal cells produce many biologicals of great value involving enzymes, hormones, growth factors, viral vaccines, or monoclonal antibodies, and many therapeutic proteins require post-translational modifications which can be only fully performed by animal cells. Insect virus, immunoregulators or whole cells, which are used for toxicological testing, can be obtained from animal cells as well as.

There is many of industrial applications of animal cells to the large scale commercial production of biologicals and they are used for in-vitro engineering of tissues like liver, skin, bone and cartilage in, as well as.

The field of animal cell biotechnology involves basic culture methods, methods for cell immortalisation, exploration of cell growth, metabolism, or productivity, molecular methods for gene transfection, as well as methods used for cultivation of genetically modified animal cells. The study of animal cells has unclosed an insight into the structure and function of cells and tissues and into biochemical and immunological processes in the cells. An issues from cell and molecular biology and biochemistry are involved in this section.

IMMORTALIZATION OF CELLS IN CULTURE

Why don't cells live forever?

Cultivating animal cells in the laboratory is an indispensable technique for cell biologists. However, most normal primary cell lines, while faithfully reproducing the phenotype of their tissue of origin, do not grow indefinitely in culture. After a series of population doublings (the number of which varies by species, cell type, and culture conditions) primary cells enter a state where they no longer divide. This state is called replicative senescence.

Replicative senescence is marked by distinct changes in cell morphology, gene expression, and metabolism and can be induced by extrinsic factors, intrinsic factors, or both. Extrinsically, irradiation, oxidative stress, or hostile cell culture environments bring about senescence by triggering the activation of various tumor suppressor proteins, including p53, Rb, and P16/INK4A. Intrinsically, the telomeric ends of chromosomes shorten with each mitotic cycle and eventually the short or uncapped ends activate these same tumor suppressor proteins, inducing senescence. Evidence suggests that this telomere shortening process serves as a counting mechanism to limit the absolute number of cell divisions and in human cells may serve as a tumour suppressor mechanism.

Why is immortalisation necessary?

Because primary cells reach senescence after a limited number of population doublings, researchers frequently need to re-establish fresh cultures from explanted tissue — a tedious process. To use the same consistent material throughout a research project, researchers need primary cells with an extended replicative capacity, or immortalised cells.

Some cells immortalise spontaneously by passing through replicative senescence and thus easily adapt to life in culture. However, these spontaneously immortalised cells invariably have unstable genotypes and are host to numerous genetic mutations, rendering them less reliable representatives of their starting tissue's phenotype.

The ideal immortalisation protocol, therefore, would produce cells that are not only capable of extended proliferation, but also possess the same genotype and tissue markers of their parental tissue.

How can cells be made immortal?

Several methods exist for immortalising mammalian cells in culture. Viral genes, including Epstein-Barr virus (EBV), Simian virus 40 (SV40) T antigen, adenovirus E1A and E1B, and human papillomavirus (HPV) E6 and E7 can induce immortalisation by a process known as viral transformation. Although the process is reliable and relatively simple, these cells may become genetically unstable (aneuploid) and lose the properties of primary cells. For the most part, these viral genes achieve immortalisation by inactivating the tumour suppressor genes that put cells into a replicative senescent state.

The preferred method to immortalise cells is through expression of the telomerase reverse transcriptase protein (TERT), particularly those cells most affected by telomere length (e.g., human). This protein is inactive in most somatic cells, but when hTERT is exogenously expressed the cells are able to maintain telomere lengths sufficient to avoid replicative senescence. Analysis of several telomerase-immortalised cell lines has verified that the cells maintain a stable genotype and retain critical phenotypic markers.

EXPLORATION ON ANIMAL MODEL FOR SENILE MEMORY DEFICITS

Objective to explore whether repeatedly administered multiple doses of scopolamine to rats is suitable for establishment of an animal model for senile memory deficits or Alzheimer disease research. Methods/14 rats were randomly divided into two groups (n=7): normal control group and scopolanmine treated group. In the experimental group, the rats received subcutaneous injection of scopolamine in a dose of 2mg/kg, twice a day for 21 consecutive days. The normal rats received the same volume of saline. Then reference memory testing in Morris water maze(MWM)was followed. Nissl staining was adopted to count the number of pyramidal cells in hippocampal ca1 and CA3 areas under light microscope. Ultrastructural changes of CA1 cells were examined by transmission electron microscopy. Results\ In the experimental group, escape latencies were statistically not significantly different from those of normal control rats on day 1, 2, 4 and 5, except day 3 in place navigation trials. In spatial probe trials, there were no significant differences between two groups. The quantity of pyramidal cells in hippocampal CA1 and CA3 areas was not significantly different between two groups. Ultrastructure of the pyramidal cells including nuclei and cytoplasmic organelles was not much changed. However, the neurons showed a decreased number of synaptic vesicles and reduced post synaptic density (PSD) in scopolamine treated rats in comparison with those of control rats. Conclusions\ Scopolamine given to rats in repeated doses may produce mild impairment of reference memory in Morris water maze. The number and ultrastructure of pyramidal neurons in CA1 and CA3 areas are not seriously changed, but some abnormalities of synapses in hippocampal CA1 cells may occur. Those findings indicate that scopolamine given in repeated doses to rats is not an ideal approach to establish an animal model for Alzheimer disease or senile memory deficits research.

GENE TRANSFER IN ANIMALS

Transfection or gene transfer in animals may be carried out at the cellular level to get transfected cells, which may be used for a variety of purposes such as production of chemicals and pharmaceutical drugs.

It may also be undertaken for basic studies involving study of structure 'and function of genes. Although many mammalian cell lines have been regularly utilized for these purposes, transfection has also been achieved successfully for the production of transgenic animals.

The improvement in livestock through transgenesis has already led to the following encouraging results:

(i) increased milk production in cattle;

(ii) increased growth rate of livestock and fish

(iii) large scale production of valuable proteins in milk, urine and blood, of livestock, enabling the use of transgenic animals as 'bioreactors' for 'molecular farming'.

In organisms like bacteria and other microbes, or even in higher plants, the uptake of genes by cells is often described by the term 'transformation'. However in animals this term has been replaced by the term 'transfection', because the term 'transformation' in animal cell culture is used to describe phenotypic alteration of cells.

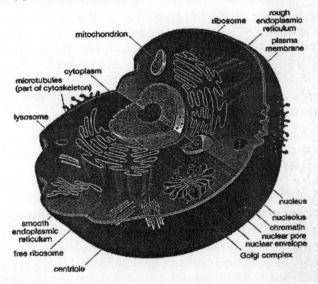

Fig. 16.1: A typical animal cell

The organisation of animal cells differences [plant vs. animal cells] are colour coded red.

The nucleus is the most conspicuous organelle found in a eukaryotic cell. It houses the cell's chromosomes and is the place where almost all DNA replication and RNA synthesis occur. The nucleus is spheroid in shape and separated from the cytoplasm by a membrane called the nuclear envelope. The nuclear envelope isolates and protects a cell's DNA from various molecules that could accidentally damage its structure or interfere with its processing. During processing, DNA is transcribed, or synthesized, into a special RNA, called mRNA. This mRNA is then transported out of the nucleus, where it is translated into a specific protein molecule. In prokaryotes, DNA processing takes place in the cytoplasm.

Ribosomes are found in both prokaryotes and eukaryotes. The ribosome is a large complex composed of many molecules, including RNAs and proteins, and is responsible for processing the genetic instructions carried by an mRNA. The process of converting an mRNA's genetic code into the exact sequence of amino acids that make up a protein is called translation. Protein synthesis is extremely important to all cells, and therefore a large number of ribosomes—sometimes hundreds or even thousands—can be found throughout a cell. Ribosomes float freely in the cytoplasm or sometimes bind to another organelle called the endoplasmic reticulum. Ribosomes are composed of one large and one small subunit, each having a different function during protein synthesis.

Mitochondria are self-replicating organelles that occur in various numbers, shapes, and sizes in the cytoplasm of all eukaryotic cells. As mentioned earlier, mitochondria contain their own genome that is separate and distinct from the nuclear genome of a cell. Mitochondria have two functionally distinct membrane systems separated by a space: the outer membrane, which surrounds the whole organelle; and the inner membrane, which is thrown into folds or shelves that project inward. These inward folds are called cristae. The number and shape of cristae in mitochondria differ, depending on the tissue and organism in which they are found, and serve to increase the surface area of the membrane.

Mitochondria play a critical role in generating energy in the eukaryotic cell, and this process involves a number of complex pathways. Let's break down each of these steps so that you can better understand how food and nutrients are turned into energy packets and water. Some of the best energy-supplying foods that we eat contain complex sugars. These complex sugars can be broken down into a less chemically complex sugar molecule called glucose. Glucose can then enter the cell through special molecules found in the membrane, called glucose transporters. Once inside the cell, glucose is broken down to make adenosine triphosphate (ATP), a form of energy.

Chloroplasts are similar to mitochondria but are found only in plants. Both organelles are surrounded by a double membrane with an intermembrane space; both have their own DNA and are involved in energy metabolism; and both have reticulations, or many foldings, filling their inner spaces. Chloroplasts convert light energy from the sun into ATP through a process called photosynthesis.

The endoplasmic reticulum (ER) is the transport network for molecules targeted for certain modifications and specific destinations, as compared to molecules that will float freely in the cytoplasm. The ER has two forms: the rough ER and the smooth ER. The rough ER is labeled as such because it has ribosomes adhering to its outer surface, whereas the smooth ER does not. Translation of the mRNA for those proteins that will either stay in the ER or be exported (moved out of the cell) occurs at the ribosomes attached to the rough ER. The smooth ER serves as the recipient for those proteins synthesised in the rough ER. Proteins to be exported are passed to the Golgi apparatus, sometimes called a Golgi body or Golgi complex, for further processing, packaging, and transport to a variety of other cellular locations.

Lysosomes and peroxisomes are often referred to as the garbage disposal system of a cell. Both organelles are somewhat spherical, bound by a single membrane, and rich in digestive enzymes, naturally occurring proteins that speed up biochemical processes. For example, lysosomes can contain more than three dozen enzymes for degrading proteins, nucleic acids, and certain sugars called polysaccharides. One function of a lysosome is to digest foreign bacteria that invade a cell. Other functions include helping to recycle receptor-proteins and other membrane components and degrading worn out organelles such as mitochondria. Lysosomes can even help repair damage to the plasma membrane by serving as a membrane patch, sealing the wound.

17
Animal Genetic Resources for Agriculture and Food Production

INTRODUCTION

Livestock play important roles in the production of food and for other purposes. The diversified use of livestock on average contributes to between 10 per cent and 50 per cent of the gross domestic product (GDP) of countries in the tropical developing world. About 70 per cent of the world's rural poor depend on livestock for their livelihood. Livestock therefore are of great socio-economic and cultural value in various societies around the world. This situation and implications for the future use of AnGR can be summarised as follows:

There is a great challenge to alleviate poverty in developing countries by producing more and safe food, especially of animal origin, against a shrinking animal genetic diversity and increased global trade. There must be a livestock revolution in the developing world to meet the projected demands of more than double the current meat and milk consumption in these countries over the next 20 years. This demand cannot only be met by an increased number of animals; increased productivity is also required.

The potential of indigenous breeds in developing countries is often inadequately documented and utilised.

The value of AnGR conservation is generally underestimated, as the current indirect values are often neglected; the future option values are yet to be accurately estimated and predicted, yet the most efficient way to sustain a breed is to continuously keep it commercially competitive or culturally viable.

Global initiatives must be locally internalised and accompanied by local activities to implement conservation programmes that increase animal productivity while maintaining the necessary genetic diversity. Previous conservation/improvement programmes have often failed. Good and simple examples that demonstrate effective breeding strategies (which take into account environmental, socio-economic and infrastructure constraints) must be developed.

Research and capacity building at all levels to improve the knowledge of indigenous and alternative AnGR in different regions of the developing world is required. The implementation of sustainable breeding strategies in the tropical developing world will be instrumental in increasing awareness of the roles of livestock and their genetic diversity.

FOOD SECURITY AND LIVESTOCK-KEYS TO POVERTY ALLEVIATION

At the dawn of the 21st century more than 1.2 billion people live in extreme poverty, while 850 million are chronically hungry and the number is rising. Most of these people are found in sub-Saharan Africa, and South and East Asia. Of the 40,000 people that die each day of malnutrition, about half are infants and children (FAO 2005a). Throughout the developing world poverty is linked to hunger and every other person in sub-Saharan Africa is considered poor, i.e. lives on less than one US dollar a day. It is estimated that 5 million children die of causes directly related to malnutrition annually.

Availability of affordable food of livestock origin would go a long way to helping alleviate this catastrophe. However, the challenge of adequately feeding people in the future is exacerbated by the fact that the global population increases by some 90 million people annually. This means that the world's farmers will have to increase their production by 50 per cent to feed about 2 billion more people by the year 2020.

The overall objective of the Millennium Development Goals (MDG) of the United Nations (UN 2004) is to reduce the proportion of people who are extremely poor and hungry by 50 per cent by the year 2015. Two of the specific MDG targets are: i) child mortality rate to be reduced by two-thirds for children under 5 years of age; and ii) environmental sustainability should be ensured. However, according to the Progress Report of the Millennium Development Goals (UN 2004), the sub-Saharan Africa, South Asia and western Asia regions still lag behind in terms of the set targets in almost all the eight goals. The incidents of extreme poverty are still very high, universal education is behind and child mortality rates remain high with no significant changes taking place. In addition, HIV/AIDS is still ravaging many populations and environmental sustainability is declining. In each of the goals, development and sustainable use of livestock, especially if targeted to the poor, provides a pathway to achieving the goals.

The increasing disparity between population growth and food production for sub-Saharan Africa is also illustrated in Fig. 17.1 (CGIAR 1999). Unless constraints to higher yields are overcome, one-third of the population in this region will not have sufficient food by 2010.

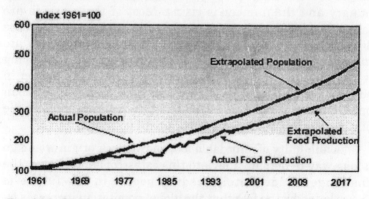

Fig. 17.1: Trends in human population growth and food production

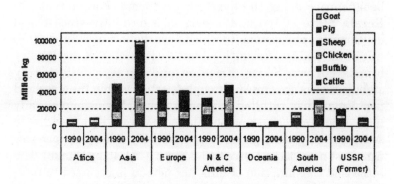

Fig. 17.2: Total world meat production by region

Enhanced food security is a key factor for poverty alleviation. The overwhelming challenge to improve the well-being of people in developing countries is thus highly dependent on the realisation of increased food production, and access to food of animal origin, in the coming decades. But if the global population increase could be curtailed at the same time then the level of increased food production required might decrease to something that may be more realistically attainable.

A study by ILRI's Livestock Policy Programme examined the food security and marketed surplus effects of intensified dairying in a peri-urban area of Addis Ababa, Ethiopia, where a market-oriented dairy production system using supplementary feed and management technologies for increased production had been introduced for smallholders. Results showed that women in households with access to crossbred cows earned nearly 7 times more dairy income than women in households with local breed cows for the same division of work, and had greater opportunities with the increased output and income. They consumed on average 22 per cent more milk and 30 per cent more calories per day and could afford 36 per cent higher food expenditures, leading to the intake of a more nutritious diet.

In India, investment in research and extension in support of crossbreeding of cattle has yielded a return rate of 55 per cent annually from the date of investment, with the primary beneficiaries being the livestock producers. For example, in Kerala State of India, 40 years of a dairy livestock-based development programme, in which a synthetic breed (Sunandini) was developed by crossing local cattle with different exotic dairy breeds (Brown Swiss, Friesian and Jersey), followed by stabilisation of the crosses through selection within the crossbred population, has resulted in great success. For example, daily milk consumption per person increased from 20 g per day to 280 g per day, through an improved daily milk yield per cow from just more than 1 litre to 6-8 litres [CS1.40 by Chako]. The increased consumption of milk per person is reported to have had a significant positive effect on child nutrition and health and huge impacts on the livelihoods of the people.

A community-based dairy goat crossbreeding and animal health-care programme in the Meru area of the Eastern highlands of Kenya has demonstrated similar examples. In the Meru area improved goat genotypes accompanied by improved husbandry practices were adopted by hitherto very poor farmers whose livelihood was well below US$ 1/head per day. Currently the same group, comprising of 3450 members, keeps improved goats each producing between 1.5 and 4 litres of milk per day. The group now produces about 3500 litres of milk daily, and is processing and packaging some of this for sale. Besides the primary producers, goat milk and meat traders and those employed along the production-to-consumption chains are also benefiting.

However, studies by Krishna et al. (2004) clearly indicated that loss of livestock assets through sales to pay for hospital bills for protracted sickness associated with HIV/AIDS and other diseases and costs associated with deaths may lead families to abject poverty. Ways out of poverty were consequently partly associated with the possession of livestock, starting with poultry and small ruminants, and at later stages with cattle. Similar results were obtained for rural communities in western Kenya and in different village types in Peru.

World animal populations increase, but not everywhere.

The distribution of livestock populations of different species by regions in 2004 is shown in Fig. 17.3.

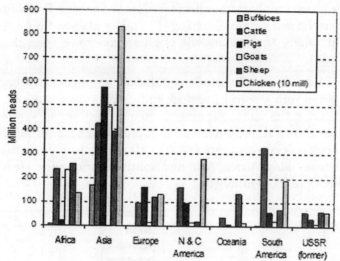

Fig. 17.3: **World livestock population by regions**

There are some striking differences, which are likely to be the result of different natural resources, climate, culture and socio-economic conditions (FAO 2005a). Whereas among the ruminants, cattle and sheep together dominate the animal populations in Asia, Africa and Oceania, the population of cattle, sheep and goats are quite similar in Europe and the former USSR. In North, Central and South America cattle dominate, while goats are primarily found in Asia (63%) and Africa, while 39 per cent of sheep are found in Asia, mostly (40%) in East Asia. The swine populations are more or less confined to Asia and the western parts of the world. Asia keeps 97 per cent of the world's buffaloes and 60 per cent of the world's swine population, of which 50 per cent are found in China. The world poultry population is estimated to be 16 billion and with the exception of Africa, Europe and the former USSR, they are fairly well distributed across regions, although Asia has the largest share (>50%,), while Africa keeps the smallest number (Figure 17.3). Of the 19 million camels in the world today, most are found in Africa.

The most remarkable changes in the past 15 years as regards species are that poultry numbers have increased by more than 50 per cent and goats by more than 30 per cent, while sheep numbers have decreased by 13 per cent. Regarding regions the most dramatic change has taken place in the former USSR, where the populations of all species have been about halved.

The different livestock population numbers have been converted into tropical livestock units (TLU) in Figure 17.4, considering the metabolic size of animals of different species. Europe shows slightly decreased animal numbers for all the livestock species, yet there is a surplus of livestock production in Europe today. Africa, Asia and South America show steady increases in TLU numbers.

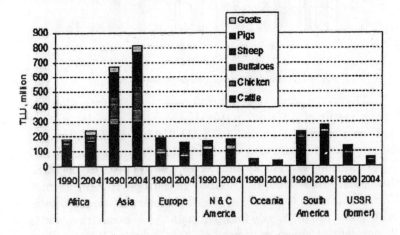

Fig. 17.4: **Trends in livestock numbers measured as total tropical livestock units (TLU) by region. Conversion factors: Buffalo and cattle 0.7; Pig 0.4; Sheep and Goat 0.1; Chicken 0.01**

When contrasting the TLU numbers with the output of food products in Figures 3 and 4 it emerges that high livestock numbers and TLU do not necessarily equate to high productivity. Neither do they reflect the overall utility functions that the various livestock play in each region. For example, whereas a cattle TLU in Africa is the same as a cattle TLU in Europe, on average the European cattle are almost 2-3 times bigger, and thus the two are not comparable from a productivity point of view. Secondly, the African/Asian animals are used for many more tasks than food production (e.g. draft, energy, social security etc.) compared to animals in temperate climates in the developed world.

To meet increasing future milk and meat demands in the developing countries, improvement in productivity will be needed. Such improvements will be realised through a combination of improved husbandry and careful utilisation of the existing livestock genotypes.

LIVESTOCK REVOLUTION UNDERWAY

Estimates of realised and projected consumption trends by the International Food Policy Research Institute (IFPRI), the Food and Agriculture Organisation of the United Nations (FAO) and the International Livestock Research Institute (ILRI) shows that production of certain food commodities will have to increase more rapidly than others in different parts of the world to meet expected demands. Whereas only marginal increases in consumption of meat and milk are expected in the developed world, increases of 114 per cent and 133 per cent respectively are projected until the year 2020 for meat and milk consumption in the developing world. The projected production increases to meet these demands in developing countries amount to 108 per cent for meat and 145 per cent for milk. The greatest (85%) increase in world meat consumption will be developing counties, with highest increases occurring in Asia, specifically East Asia. Also, more than 90 per cent of the world's predicted 60 per cent increase in milk consumption will occur in Asia, mainly South Asia (FAO 2005a). However, for the next 10 to 25 years, minimal growth will take place in the overall global consumption of these two livestock products.

The demands for increased animal products are higher than for cereals because of changing consumption patterns following urbanisation, population growth and projected income growth. Diets with more high-value protein and micronutrients will improve human health and the livelihood of many poor people. The implications of increased food production and changed diets of billions of people may be dramatic in the next few decades and could improve the well-being of many rural poor as both consumers and producers.

In contrast to the familiar Green Revolution that started in plant production 30 years ago, a livestock revolution is just underway to meet the increase in demand for food of animal origin. Such a revolution assumes a wise use of natural resources, including animal and plant genetic resources, in order to be realised. The challenge is how to take advantage of prevailing trends for the benefit of the rural and sub-urban poor livestock keepers in developing countries rather the more industrialised production in other parts of the world. Already predictions are that unless major improvement in productivity occurs, East Asia and Africa will increasingly remain net importers of meat and milk products. For cereals, milk and dairy products, South Asia, Africa and East Asia will increasingly become net exporters of cereals. More than 70 per cent of the predicted increase in the world's meat consumption will be in form of pork and poultry, most of which will be produced under intensive industrial production, partly explaining the predicted trends in inter-regional cereal trade.

The higher pace of industrialisation will continue, especially for pig and poultry production. This process is predicted to drive the small producers out of the ever-increasing competitive global market, for both economic and biological reasons. The benefits derived from economies of scale, leading to better resilience against disasters and calamities such as the ongoing bird flu outbreaks and favourable domestic trade support and policy environments may further favour industrialised livestock production systems in the future.

Although mixed crop-livestock production systems will persist in the foreseeable future, higher levels of intensification will be required, with increased use of livestock genotypes that are likely to respond better to the changes in production systems. Consequently, small-scale mixed crop-livestock production systems will eventually be confined to more remote areas, with poverty persisting and livestock playing a more central survival role and a key first step out of poverty. Under such conditions livestock on their own are unlikely to create overwhelming riches to their keepers.

ILRI, in its strategy to 2010, has identified activities in livestock research and development (R&D) for developing countries, which focus on poverty reduction, food and nutritional security and environment and human health. It includes a substantial programme on characterisation of indigenous AnGR and development of strategies for sustainable utilisation of the diversity in livestock species, which assumes increased productivity, to improve the livelihoods of people in developing countries. Equally important, innovative ways must be sought to secure market access for the livestock products of developing countries. If this is not achieved the globalisation in trade of agricultural products is going to wipe out less competitive production systems in favour of products from industrialised ones in other parts of the world.

DIVERSIFIED USE OF LIVESTOCK

Domestic animals have, for more than 10 thousand years, contributed to human needs for food and agricultural products such as meat, dairy products, eggs, fibre and leather, draft power and transport, manure to fertilise crops and for fuel. Livestock also play an important economic role as capital and for social security.

The value of livestock has also been clearly demonstrated for soil nutrient management, especially in soils in rapidly intensifying crop-livestock systems and in those already intensified. Integration of livestock into crop systems enhances smallholder farm productivity and profitability.

The multiple uses of livestock also include their cultural roles in many societies. Hence, the use of animal resources varies considerably between various parts of the world as the social, environmental and other conditions for animal production differ enormously.

Currently, an estimated 30-40 per cent of the world's total agricultural output is produced by its variety of livestock. In some parts of the world, including some parts of Africa, where intensive mixed livestock-crop systems are practised, as much as 70-80 per cent of the farm income is from livestock. In such systems, much of the crops produced are fed to livestock and converted to high quality food for human consumption.

DIVERSITY IN ANIMAL GENETIC RESOURCES INVALUABLE FOR FUTURE DEVELOPMENTS

The consistent contribution over thousands of years of animal production to human needs under different environmental conditions as diverse as, arctic and tropical, maritime and mountain, humid and arid semi-desert ecozones, stems from the development of some 4000-5000 breeds of different species. Of these, about 70 per cent are found in the tropical developing world [DAD-IS; DAGRIS;]. They have been domesticated from about 40 wild animal species according to different needs and uses under the variable environments that have covered the world over time. The adaptation of different species and breeds to a broad range of environments provides the necessary variability that offers opportunities to meet the increased future demands for food and provide flexibility to respond to changed markets and needs [Breed information].

CONSIDERABLE GENETIC VARIATION AMONG BREEDS

The diversity among breeds is known to contribute about half of the genetic variation found among animals within species, while the other half is attributed to genetic variation within breeds. The variation within breeds is less vulnerable to loss but breeds are easily irretrievably lost when they are considered to be commercially non-competitive. That is why the maintenance of local breeds is of great importance for the maintenance of genetic diversity. However, it may not be possible to maintain all breeds forever, especially if they are not competitive enough, all values considered. The definition of a breed is somewhat arbitrary and has, throughout history, allowed for some dynamics. Some breeds are disappearing or have disappeared, while others have been formed [Breed information]. Such changes have been possible and necessary as part of the evolution and the dynamics that the variability of the genetic resources allows for interaction with environmental changes.

In the absence of appropriate breed characterisation, breed attributes and genes that are potentially beneficial in the future may not be saved. Instead, some breeds are condemned to extinction and in the process, some of the good genes that they may have possessed disappear with them, never to be recovered. However, well planned crossbreeding systems could help save the desirable genes, even when the livestock breeds that once posses such genes are lost. The successive development of a synthetic breed is a typical example on how valuable genes can be saved for the future.

WHY WORRY ABOUT LOSS IN GENETIC DIVERSITY?

Genetic improvement of animal populations is dependent on the existence of genetic variation. Such variation exists between species, between breeds within species and among animals within breeds. As species and breeds are adapted to certain environments, through centuries or thousands of years of natural and artificial selection, it may be difficult to restore such genetic variation that may still be desired, but that has been lost by breed replacements in certain regions or environments. The continuous loss of breeds and genetic diversity is usually fuelled by short-sighted and restricted genetic and socio-economic considerations. The real long-term values, including ecological effects, may not have been taken into account. Also not usually considered are future changes that may have an impact on the needs for variable genetic resources. The irreversible losses of genetic diversity therefore, reduce our opportunities for future developments. That is why it is imperative to critically consider both the present and future breeding programmes of all species and breeds in relation to environmental and economic developments and needs.

The distribution of species by world regions may lead to the conclusion that ruminants, which today have the largest world coverage and are represented by a large number of breeds that are adapted to different environments, would have the best opportunities to adapt to future environmental changes. Similarly, populations confined to a few regions or specialised production systems are more vulnerable to changes in production or economic systems in those regions. Such effects may dramatically reduce the genetic diversity and our future opportunities for development of efficient animal food production under variable conditions. The importance of the Asian region, and especially China, for conservation of a variety of indigenous pig breeds is extremely high, as these breeds are not found elsewhere.

NEW APPROACHES NEEDED FOR SUSTAINABLE LIVESTOCK IMPROVEMENT

The awareness of the demands for increased productivity has not been lacking. In fact, many attempts have been made to genetically improve livestock in the tropics. Although it should be recognised that improved livestock have been produced or successfully introduced in favourable areas of the tropics, e.g. in some highland areas, in maritime climates and in relatively intense peri-urban production systems, many attempts have failed. At least three primary reasons could be seen for these failures:

Due to lack of domestic resources and enough trained staff with an animal breeding background, people from developed countries have usually been responsible for conducting improvement programmes. As a consequence of this lack of 'indigenous' knowledge, sophisticated methods, e.g. use of artificial insemination and progeny testing, have often been inappropriately applied, neglecting the necessary infrastructure.

The introduction of crossbreeding with temperate high yielding breeds without a long-term plan on how to maintain either a suitable level of 'upgrading', or how to maintain the pure breeds for future use in crossbreeding has been another reason. Upgrading to a level that is too high has generally led to animals without resistance to withstand environmental stress. However, there are examples of successful breed replacements in parts of India and Africa, including the highlands of Kenya. Furthermore, use of intermediate crossbred cattle based on introduced breeds has been successfully demonstrated in Brazil and is one way of combining diverse genetic attributes of the different breeds, so long as an organised crossbreeding programme is followed.

The lack of analysis of the different roles of livestock in each specific area, usually leads to falsely defined breeding objectives and underrating the potentials of various indigenous breeds of livestock.

New approaches must better consider the potential of indigenous livestock breeds sometimes in crossbreeding with suitable exotic breeds, and realistic ways of improving them in the context of environmental and socio-economic demands and within the resources available. For this purpose there is a great need to characterise the indigenous livestock breeds and their crosses to determine which are the most suitable ones for further improvement and to implement simplified, but yet effective, breeding programmes. Such programmes could be based on nucleus herds of pure and crossbred animals from which specified genotypes or semen can be widely disseminated to livestock herds.

18

Animal Monoclonal Antibody

INTRODUCTION

Monoclonal antibodies, following the development of this technology in 1975, have become an extraordinarily important resource for medical research, diagnosis, therapy, and basic science. In recognising the overwhelming importance of this technology to modern medical science, it is also important to recognise that monoclonal antibody production is still largely dependent on the use of experimental animals.

In November 1997, the Office for Protection from Research Risks (OPRR) at the National Institutes of Health, forwarded a letter to all Public Health Service (PHS) awardee institutions and Institutional Animal Care and Use Committees (IACUCs) on avoiding or minimising discomfort, distress, and pain in the care and use of animals for the production of monoclonal antibodies using mouse ascites method since there is evidence that the ascites method of monoclonal antibody production causes discomfort, distress, or pain. Accordingly, the IACUCs at all Public Health Service awardee institutions must critically evaluate the proposed use of the ascites method. The Committee must determine that:

(a) the proposed use is scientifically justified;

(b) methods that avoid or minimise discomfort, distress, and pain (including in vitro methods) have been considered; and

(c) the latter have been found unsuitable. Fulfillment of this three-part IACUC responsibility, with appropriate documentation, is considered central to an institution's compliance with its Animal Welfare Assurance and the PHS Policy.

In keeping with the need to minimise both the use and the discomfort and pain inflicted on experimental animals, the following considerations and recommendations in the application of monoclonal antibody technology are proposed.

MONOCLONAL ANTIBODY TECHNOLOGY

Monoclonal antibody technology relies on experimental animals for two basic steps. The first step involves immunising mice. Although there are alternatives to this step (either by in vitro immunisation steps or using recombinant libraries), these alternatives have not proven to provide the same level of efficiency, specificity, or affinity of antibody available from traditional immunisation protocols. In addition, with the advent of alternatives to Freund's adjuvant, less than minimal or slight pain or discomfort accompanies this step and therefore there is no compelling reason to consider alternatives at this time. Careful consideration of the total number of required animals should be the investigators main concern in justification of this step .

A second common step that uses animals comes in antibody production. In order to produce large quantities of monoclonal antibody, it has been traditional to grow hybridoma cell lines as an ascites tumour in vivo. In general, this procedure produces antibody titers approximately 1000-10,000-fold higher than those obtained in tissue culture. Since 1-10 ml of ascites fluid containing 1-5 mg/ml specific antibody can be obtained per animal while hybridoma cells in culture produce only 0.5-5 µg/ml, one mouse can produce antibody equivalent to between 1 and 100 litres of tissue culture fluid. The main advantages of ascites are the extremely high yield of antibody and that the method is not excessively labour-intensive. However, its main disadvantage is the potential pain and discomfort caused to animals, due to painful peritonitis, abdominal tension, and infiltratively-growing tumours. It should also be noted that the monoclonal antibody produced by this method is contaminated by endogenous immunoglobulin and has the potential for contamination by viruses or bioreactive cytokines that may interfere with later use.

There are in vitro alternatives to monoclonal antibody production by ascites. These include standard static or agitated suspension cell cultures, membrane-based and matrix-based culture systems, and high cell-density bioreactors. The disadvantages to these systems include their substantially greater effort and higher labor costs, increased costs due to the components of tissue culture media, and the poor growth and/or antibody secretion of some hybridoma lines in vitro. In particular, high cell-density bioreactors are probably beyond the capability of most laboratories due to the high initial and ongoing material and labor costs and the specialised expertise required.

In light of the above considerations, the New York University IACUC requests that investigators consider the following recommendations in designing monoclonal antibody production.

In general, in vitro methods for monoclonal antibody production are considered standard and accepted practice. The use of the ascites method requires rigorous and well-documented justification. Justifications based solely on cost or convenience will not be considered adequate.

In addition, the NYU IACUC strongly urges the Medical Centre to establish a monoclonal antibody core facility with high-density cell culture bioreactor capability that would obviate any need for continued use of the ascites method.

Most applications require only small quantities and low concentrations of antibodies. Examples of common applications requiring only small quantities of antibody include immunoblots, immunoprecipitations, immunocytochemistry, flow cytometry, and small-scale affinity columns. Such applications are easily accommodated by use of unpurified tissue culture supernatants or by monoclonal antibodies purified from tissue culture supernatants. For purposes requiring up to approximately 10-50 mg of antibody, standard tissue culture methods involving growth of up to 50 litres of hybridoma cells should be considered the method of choice. Proposed use of ascites for applications of this kind would require specific justification (e.g., use of a hybridoma with unfavorable growth characteristics in vitro) in addition to the considerations described.

Small-scale membrane-based culture systems are available that facilitate production of monoclonal antibodies in the range of 10-100 mg per culture in 5-30 days. These systems are relatively inexpensive and do not require specialised facilities or expertise. Investigators with applications that require up to 1 g of monoclonal antibody should consider the use of this in vitro alternative for production. Use of ascites production for such applications would require specific justification demonstrating that special circumstances warrant use of the in vivo method. In addition to the considerations described, justification must demonstrate that the disadvantages of ascites production (including pain and discomfort to the animals and potential in vivo contamination) are outweighed by the specific requirements of the individual project.

PRODUCTION OF MONOCLONAL ANTIBODY

Production of greater than 1 g of monoclonal antibody by in vitro methods is probably outside the current capabilities of most laboratories. For these applications, ascites production is presently the only alternative for in-house production. Approval of such projects will require adequate justification for the required amounts of antibody. In addition, the investigator must demonstrate the required expertise for working with tumour-bearing mice, including an adequate daily monitoring system to insure that animals do not experience unnecessary pain or discomfort. Assurance must be made that no animal will be allowed to develop tumours larger than 20 per cent of host body weight. Ascites fluid must be harvested on a single occasion only, either under terminal anaesthesia or post mortem. Animals must be killed without delay if they show more than mild distress, overt tumor deposits or spread, or significant dehydration or cachexia.

The specific guidelines for consideration by Principal Investigators when developing animal study proposals and for Animal Care and Use Committees when reviewing proposals involving the mouse ascites method are:

(a) The volume of the priming agent should be reduced to as small a volume as necessary to elicit the growth of ascitic tumours and at the same time reduce the potential for distress caused by the irritant properties of the priming agent. Although 0.5 ml Pristane has been standard for adult mice, 0.1-0.2 ml has been found to be as effective for many hybridomas;

(b) The time interval between priming and inoculation of hybridoma cells as well as the number of cells in the inoculum are determined empirically. Inocula range from 105-107 cells in volumes of 0.1-0.5 ml and are usually administered 10-14 days after priming. Generally, very high concentrations are associated with greater mortality and concentrations $< 1 \times 105$ cells elicit fewer ascitic tumors and these tend to have a smaller volume yield. Cell suspensions should be prepared under sterile conditions in physiological solutions;

(c) Hybridomas should be MAP (mouse antibody production) or PCR tested before introduction into the animal host to prevent potential transmission of infectious agents from contaminated cell lines into facility mouse colonies and possibly to humans handling the animals;

(d) Animals should be monitored at least once daily, seven days a week by personnel familiar with clinical signs associated with ascites production and circulatory shock;

(e) Ascites pressure should be relieved before abdominal distension is great enough to cause discomfort or interfere with normal activity. Manual restraint or anesthesia may be used for tapping. Aseptic technique should be used in withdrawing ascitic fluid. The smallest needle possible that allows for good flow.

In accordance with the Animal Welfare Assurance and the Public Health Service Policy, and to appropriately document that investigators proposing the use of monoclonal antibodies have considered alternatives to minimise discomfort, distress, and pain, in future please refer to these recommendations before submitting protocols to the IACUC.

HYBRIDOMA CELL PRODUCTION

Monoclonal antibodies are typically made by fusing myeloma cells with the spleen cells from a mouse that has been immunised with the desired antigen. However, recent advances have allowed the use of rabbit B-cells. Polyethylene glycol is used to fuse adjacent plasma membranes, but the success rate is low so a selective medium is used in which only fused cells can grow. This is because myeloma cells have lost the ability to synthesize hypoxanthine-guanine-phosphoribosyl transferase (HGPRT).

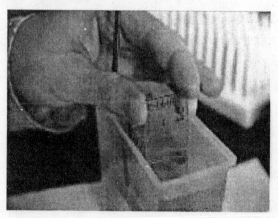

Fig. 18.1: Lab technician bathing prepared slides in a solution. This technician prepares slides of monoclonal antibodies for researchers. The cells shown are labelling human breast cancer

This enzyme enables cells to synthesize purines using an extracellular source of hypoxanthine as a precursor. Ordinarily, the absence of HGPRT is not a problem for the cell because cells have an alternate biochemical pathway that they can use to synthesize purines. However, when cells are exposed to aminopterin (a folic acid analogue), they are unable to use this other, rescue, pathway and are now fully dependent on HGPRT for survival. The selective culture medium is called HAT medium because it contains Hypoxanthine, Aminopterin, and Thymidine. This medium is selective for fused (hybridoma) cells because unfused myeloma cells cannot grow because they lack HGPRT. Unfused normal spleen cells cannot grow indefinitely because of their limited life span. However, hybridoma cells are able to grow indefinitely because the spleen cell partner supplies HGPRT and the myeloma partner is immortal because it is a cancer cell. The fused hybrid cells are called hybridomas, and since they are derived from cancer cells, are immortal and can be grown indefinitely.

Fig. 18.2: **Technician hand-filling wells with a liquid for a research test. This test involves preparation of cultures in which hybrids are shown in large quantities to produce desired antibody. This is effected by fusing myeloma cell and mouse lymphocyte to form a hybrid cell (hybridoma)**

Animal Monoclonal Antibody

This mixture of cells is then diluted and clones are grown from single parent cells. The antibodies secreted by the different clones are then tested for their ability to bind to the antigen (for example with a test such as EIA or Antigen Microarray Assay) or immuno-dot blot, and the most productive and stable clone is then grown in culture medium to a high volume. When the hybridoma cells are injected in mice (in the peritoneal cavity, the gut), they produce tumours containing an antibody-rich fluid called ascites fluid.

The medium must be enriched during selection to further favour hybridoma growth. This can be achieved by the use of a layer of feeder fibrocyte cells or supplement medium such as briclone. Production in cell culture is usually preferred as the ascites technique is painful to the animal and if replacement techniques exist, this method is considered unethical.

RECOMBINANT

The production of recombinant monoclonal antibodies involves technologies, referred to as repertoire cloning or phage display/yeast display. Recombinant antibody engineering involves the use of viruses or yeast to create antibodies, rather than mice. These techniques rely on rapid cloning of immunoglobulin gene segments to create libraries of antibodies with slightly different amino acid sequences from which antibodies with desired specificities can be selected. These techniques can be used to enhance the specificity with which antibodies recognise antigens, their stability in various environmental conditions, their therapeutic efficacy, and their detectability in diagnostic applications. Fermentation chambers have been used to produce these antibodies on a large scale.

APPLICATIONS

Once monoclonal antibodies for a given substance have been produced, they can be used to detect the presence and quantity of this substance, for instance in a Western blot test (to detect a protein on a membrane) or an immunofluorescence test (to detect a substance in a cell). They are also very useful in immunohistochemistry which detect antigen in fixed tissue sections. Monoclonal antibodies can also be used to purify a substance with techniques called immunoprecipitation and affinity chromatography.

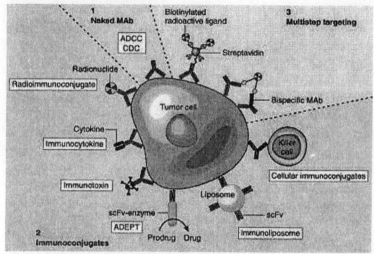

Fig. 18.3: Monoclonal antibodies for cancer. ADEPT, antibody directed enzyme prodrug therapy; ADCC, antibody dependent cell-mediated cytotoxicity; CDC, complement dependent cytotoxicity, MAb, monoclonal antibody; scFv, single-chain Fv fragment

MONOCLONAL ANTIBODIES FOR CANCER TREATMENT

One possible treatment for cancer involves monoclonal antibodies that bind only to cancer cell-specific antigens and induce an immunological response against the target cancer cell. Such mAb could also be modified for delivery of a toxin, radioisotope, cytokine or other active conjugate; it is also possible to design bispecific antibodies that can bind with their Fab regions both to target antigen and to a conjugate or effector cell. In fact, every intact antibody can bind to cell receptors or other proteins with its Fc region.

One problem in medical applications is that the standard procedure of producing monoclonal antibodies yields mouse antibodies. Although murine antibodies are very similar to human ones there are differences. The human immune system hence recognizes mouse antibodies as foreign, rapidly removing them from circulation and causing systemic inflammatory effects.

A solution to this problem would be to generate human antibodies directly from humans. However, this is not easy, primarily because it is generally not seen as ethical to challenge humans with antigen in order to produce antibody; the ethics of doing the same to non-humans is a matter of debate. Furthermore, it is not easy to generate human antibodies against human tissues.

Various approaches using recombinant DNA technology to overcome this problem have been tried since the late 1980s. In one approach, one takes the DNA that encodes the binding portion of monoclonal mouse antibodies and merges it with human antibody producing DNA. One then uses mammalian cell cultures to express this DNA and produce these half-mouse and half-human antibodies. (Bacteria cannot be used for this purpose, since they cannot produce this kind of glycoprotein.) Depending on how big a part of the mouse antibody is used, one talks about chimeric antibodies or humanised antibodies. Another approach involves mice genetically engineered to produce more human-like antibodies. Monoclonal antibodies have been generated and approved to treat: cancer, cardiovascular disease, inflammatory diseases, macular degeneration, transplant rejection, multiple sclerosis, and viral infection.

TYPES OF MONOCLONAL ANTIBODIES

Infliximab

Infliximab (brand name Remicade) is a drug used to treat autoimmune disorders. Infliximab is known as a "chimeric monoclonal antibody" (the term "chimeric" refers to the use of both mouse (murine) and human components of the drug i.e. murine binding VK and VH domains and human constant Fc domains). The drug blocks the action of TNF? (tumour necrosis factor alpha) by binding to it and preventing it from signaling the receptors for TNF? on the surface of cells. TNF? is one of the key cytokines that triggers and sustains the inflammation response.

BASILIXIMAB

Basiliximab (Simulect) is a chimeric mouse-human monoclonal antibody to the IL-2R? receptor of T cells. It is used to prevent rejection in organ transplantation, especially in kidney transplants. It is a Novartis Pharmaceuticals product and was approved by the Food and Drug Administration (FDA) in 1998.

Fig. 18.4

It is given in two doses, the first within 2 hours of the start of the transplant operat n and the second 4 days after the transpl nt. These saturate the receptors and prevent T cells from replication and also from activating the B cells, which are responsible for the production of antibodies, which would bind to the transplanted organ and stimulate an immune response against the transplant.

ABCIXIMAB

Abciximab (previously known as c7E3 Fab), manufactured by Centocor and distributed by Eli Lilly under the trade name ReoPro, is a platelet aggregation inhibitor mainly used during and after coronary artery procedures like angioplasty to prevent platelets from sticking together and causing thrombus (blood clot) formation within the coronary artery. Its mechanism of action is inhibition of glycoprotein IIb/IIIa.[citation needed]

While Abciximab has a short plasma half life, due to its strong affinity for its receptor on the platelets, it may occupy some receptors for weeks. In practice, platelet aggregation gradually returns to normal about 24 to 48 hours after discontinuation of the drug.[citation needed]

Abciximab is made from the Fab fragments of an immunoglobulin that targets the glycoprotein IIb/IIIa receptor on the platelet membrane.[citation needed]

DACLIZUMAB

Daclizumab (Zenapax) is a therapeutic humanised monoclonal antibody to the alpha subunit of the IL-2 receptor of T cells. It is used to prevent rejection in organ transplantation, especially in kidney transplants.

It is given in multiple doses, the first 1 hour before the transplant operation and 5 further doses given at two week intervals after the transplant. These saturate the receptors and prevent T cell activation and thus prevent formation of antibodies against the transplant.

Like the similar drug basiliximab, daclizumab reduces the incidence and severity of acute rejection in kidney transplantation without increasing the incidence of opportunistic infections.

Daclizumab usage may also be indicated in place of a calcineurin-inhibitor (ciclosporin or tacrolimus) during the early phase after kidney transplantation, when the kidney is recovering and vulnerable to calcineurin-inhibitor toxicity. This has been shown to be beneficial in non-heart beating donor kidney transplantation.

In the United Kingdom, the National Institute for Health and Clinical Excellence has recommended its use be considered for all kidney transplant recipients.

GEMTUZUMAB OZOGAMICIN

Gemtuzumab ozogamicin (marketed by Wyeth as Mylotarg) is a monoclonal antibody used to treat acute myelogenous leukemia.

It is a monoclonal antibody to CD33 linked to a cytotoxic agent, calicheamicin. CD33 is expressed in most leukemic blast cells but also in normal hematopoietic cells the intensity diminishing with maturation.stem cells. When given to patients in first relapse, 15 per cent of patients achieve a complete remission. In the United States, it was approved by the FDA for use in patients over the age of 60 with relapsed AML who are not considered candidates for standard chemotherapy.

A notable complication of therapy with gemtuzumab is the increased risk of veno-occlusive disease in the absence of bone marrow transplantation. The onset of VOD can also be delayed and can occur at increased frequency following bone marrow transplantation.

Common side effects of administration include shivering, nausea, and fever. Serious side effects include severe myelosuppression (found in 98 per cent of patients), disorder of the respiratory system, tumour lysis syndrome, and Immune hypersensitivity syndrome.

ALEMTUZUMAB

Alemtuzumab (marketed as Campath, MabCampath or Campath-1H) is a monoclonal antibody used in the treatment of chronic lymphocytic leukemia (CLL) and T-cell lymphoma.

Alemtuzumab targets CD52, a protein present on the surface of mature lymphocytes, but not on the stem cells from which these lymphocytes were derived. It is used as second line therapy for CLL. It was approved by the Food and Drug Administration for patients who have been treated with alkylating agents and who have failed fludarabine therapy.

A significant complication of therapy with alemtuzumab is that it significantly increases the risk for opportunistic infections, in particular, reactivation of cytomegalovirus.

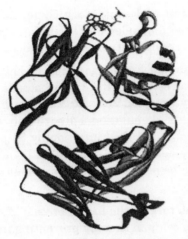

Fig. 18.5

Alemtuzumab is also used in some conditioning regimens for bone marrow transplantation and kidney transplantation. It is also used under clinical trial protocols for treatment of some autoimmune diseases, such as multiple sclerosis.

RITUXIMAB

Rituximab, sold under the trade names Rituxan and MabThera, is a chimeric monoclonal antibody used in the treatment of B cell non-Hodgkin's lymphoma, B cell leukemia, and some autoimmune disorders.

Rituximab was developed by IDEC Pharmaceuticals and initially approved by the FDA in 1997 for lymphoma that was refractory to other chemotherapy regimens. The original approval followed the availability of the McLaughlin et al study data. It now is standard therapy in the initial treatment of aggressive lymphomas (e.g. diffuse large B cell lymphoma) in combination with CHOP chemotherapy. It is currently co-marketed by Biogen Idec and Genentech in the U.S. market and Roche in the EU.

PALIVIZUMAB

Palivizumab (brand name Synagis) is a monoclonal antibody produced by recombinant DNA technology. It is used in the prevention of Respiratory Syncytial Virus (RSV) infections. It is recommended for certain infants that are high-risk (because of prematurity or other medical problems).

Palivizumab is a humanised monoclonal antibody (IgG) directed against an epitope in the A antigenic site of the F protein of the Respiratory Syncytial Virus (RSV). In two Phase III clinical trials in the pediatric population, Palivizumab reduced the risk of hospitalisation due to RSV infection by 55 per cent and 45 per cent. Palivizumab is dosed once a month via intramuscular (IM) injection, to be administered throughout the duration of the RSV season.

Palivizumab targets the fusion protein of RSV, inhibiting its entry into the cell and thereby preventing infection.

TRASTUZUMAB

Trastuzumab (more commonly known under the trade name Herceptin) is a humanised monoclonal antibody that acts on the HER2/neu (erbB2) receptor. Trastuzumab's principal use is as an anti-cancer therapy in breast cancer in patients whose tumours over-express (that is, "produce more than the usual amount of") this receptor. Trastuzumab is administered either once a week or once every three weeks intravenously for 30 to 90 minutes.

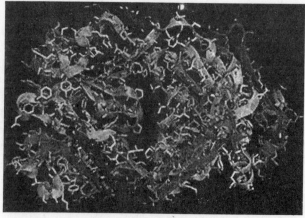

Fig. 18.6

Etanercept (Enbrel) is a recombinant-DNA drug made by combining two proteins (a fusion protein). It links human soluble TNF receptor to the Fc component of human immunoglobulin G1 (IgG1).

It is a large molecule, with a molecular weight of 150 kDa., that binds to TNF? and decreases its role in disorders involving excess inflammation in humans and other animals, including autoimmune diseases such as ankylosing spondylitis, juvenile rheumatoid arthritis, psoriasis, psoriatic arthritis, rheumatoid arthritis, and, potentially, in a variety of other disorders mediated by excess TNF?.

This therapeutic potential is based on the fact that TNF-alpha is the "master regulator" (as coined by Marc Feldmann, PhD, and Ravinder N. Maini BCh, recipients of the 2003 Lasker Award for their anti-TNF research in rheumatoid arthritis) of the inflammatory response in many organ systems.

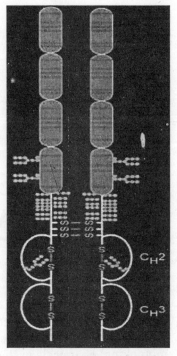

Fig. 18.7

In the United States and the United Kingdom, etanercept is co-marketed by Amgen and Wyeth under the trade name Enbrel in two separate formulations, one in powder form, the other as a pre-mixed liquid.

19

Morphological Abnormalities in Animals

INTRODUCTION

Due to the alarming numbers of animals from many species that have been found with gross morphological abnormalities, the topic of endocrine disruption, or "hormone-mimicking" chemicals in the environment, has attracted great attention in recent years. Animals as diverse as mammalian species like the Florida panther, avian species, and even reptilian species such as alligators have all been reported with defects, particularly in reproductive organs. Frog populations across the country and here in Vermont have been rapidly declining, and the numbers of deformities being reported are also on the rise. Because exposure to these chemicals has such serious implication for both wildlife populations and for human health, research directed at identifying endocrine disrupting chemicals and their biological effects is at the foreground of active research efforts.

ENDOCRINE DISRUPTING COMPOUNDS (EDCS)

Endocrine Disrupting Compounds (EDCs) are synthetic compounds found in pesticides, herbicides, nonionic surfactants, environmental pollutants, and common plastics, as well as natural compounds derived from plants that have deleterious effects on the development of a wide range of species by disrupting hormone-sensitive processes. Many studies have shown that exposure to EDCs during early development induces abnormalities in peripheral reproductive organs and in reproductive behaviors, as well as disruption of limb development. In addition to causing infertility and fetal malformations EDCs have also been shown to act as carcinogens in mammalian populations.

Studies demonstrating that early EDC exposure leads to aberrant reproductive behaviours in adult life suggest that these compounds affect not only the formation of peripheral reproductive structures, but also the developing central nervous system (CNS). Endogenous hormones (i.e., gonadal steroids: androgens, estrogens and progestins) are known to have significant and widespread effects on the development of the nervous system providing a myriad of potential targets for disruption by EDCs. Determining how EDCs alter nervous system development, however, is a complicated affair since the endpoints of assessment for nervous system abnormalities are often less easily defined than with assessments for limb malformation or tumour formation.

Moreover, the EPA has identified over 87,000 chemicals that need to be screened for potential EDC effects). This overwhelming number of chemicals, coupled with the fact that effects on nervous system development may be both hard to categorise (changes in cognitive function or affect) and variable (different in individuals with different genetic backgrounds), makes for a daunting task. Finally, assessments of which EDCs pose a health danger and at what level are controversial and, at this time, unresolved. For example, it has been estimated that ~60 per cent of the greater than 300,000 tons of alkylphenol polyethoxylates end up in the water supply each year. At the source (e.g., sewage treatment plants, mills, and factories), these compounds are detected at ~0.1 to 1 mg/litre or 10-6 M. However, the metabolites of alkylphenol polyethoxylates are highly stable and accumulate in sediment and sludge at concentrations that exceed those of the parent EDC). Compounding this physical accumulation, EDCs bioaccumulate in fatty animal tissues. TCDD (dioxin), a contaminant that derives from the commercial preparation of certain herbicides , has been measured at up to 6 ppt per mL serum in human adults . The EDC metabolites that are consumed by bottom feeding fish become increasingly concentrated as fish become eaten by birds and so on up the food chain.

What are the concentrations of EDCs required to elicit significant biological effects? Studies of how EDCs can activate estrogen-sensitive ("reporter") genes in isolated cells in culture indicate that concentrations from 10^{-8} to 10^{-5} M induce significant effects . Hypothalamic neurons maintained in dissociated cell culture are highly sensitive to EDCs, and significant effects in neurotransmitter uptake can be elicited by concentrations of alkylphenol polyethoxylates as low as 10^{-11} M . In addition, it should be noted that assays of cultured cells or reporter gene constructs do not take into account a number of critical parameters including metabolism of EDCs, bioaccumulation, or bioavailability (that is whether they are free or bound to proteins in serum that preclude them from having a biological effect at intracellular steroid receptors). Moreover, these simple assays do not take into account steroid-receptor independent mechanisms of action, or cell-cell interactions that may induce effects in an intact animal that would not be evident in cultured cells. Finally, differences in genetic background and developmental age (see below) will impose significant differences in the ability of EDCs to elicit biological effects.

CELLULAR AND MOLECULAR MECHANISM UNDERLYING EDC EFFECTS

EDCs cause adverse effects by interfering with endogenous hormonal signaling mechanisms . Endogenous steroid hormones, such as testosterone or the estrogen 17þ-estradiol, are small hydrophobic molecules that easily diffuse through the plasma membrane of a cell and into the cytoplasm where they then bind to a specific target receptor (androgen or estrogen receptors). This steroid/receptor complex then travels to the nucleus where its actions ultimately alter the biological response of the targeted cells and the organism. The overall mechanism of these steroid effects is relatively well understood. Once in the nucleus, the steroid/receptor complex directly regulates the expression, or "transcription," of specific genes by binding to discrete regulatory sequences of these genes called hormone response elements. These steroid-dependent changes in gene expression result in changes in the synthesis, or "translation," of specific proteins. It is the actions of these proteins which determine the biological responses of the targeted cells, and therefore of the organism.

EDCs could have potential deleterious actions in either of two ways: if they interfere with the normal activation of a specific receptor by the natural hormone (i.e., act as an antagonist); or if they act in the same way as the endogenous hormone (i.e., act as a hormone mimic or agonist), but at an inappropriate developmental time, or if they are present at the wrong concentration. Recent studies indicate that both mechanisms come into play. Many of the EDCs are known to exert their effects by acting as weak estrogens . For example, alkylphenolic polyethoxylates were shown to bind to estrogen receptors over twenty years ago , and more recent studies have demonstrated that putative EDCs can mimic the molecular effects of estrogen. Specifically, EDCs produce transcriptional activation of reporter gene constructs containing consensus estrogen response elements in cell lines. In the past few years, however, it has also become clear that a number of EDCs exert their effects not as weak estrogens, but rather by acting as anti-androgens. Specifically, the fungicide, vinclozolin, , the ubiquitous pesticides 1,1,1-trichloro-2, 2-bis (p-chlorophenyl) ethane (DDT) and its major metabolite, p, p'-dichlorodiphenyldichloro-ethylene (p, p'-DDE) , and bisphenol A and butyl benzyl phthalate all interfere with androgen-dependent activation of reporter constructs and alter sexual differentiation in male rodents in a manner consistent with anti-androgenic activity.

CRITICAL PERIODS IN DEVELOPMENT

For most biological processes, but especially those related to hormone effects on the nervous system, there is incontrovertible data indicating that neural processes are significantly more susceptible to steroid effects during embryonic and early postnatal development than in adulthood . In particular, it is known that naturally-occurring hormones can induce significant changes in neurogenesis (the birth of nerve cells or neurons), neuronal survival, neuronal migration, the connections neurons make with one another, as well in the expression of specific proteins that determine neuronal function during these early developmental "critical periods." Moreover, these changes are permanent and do not require continued presence of high levels of hormones. As development proceeds, however, many facets of this hormone-sensitivity are lost, and the adult brain is far less malleable with respect to these "organisational" actions of steroid hormones . While there is far less known about the organizational actions of EDCs, several studies suggest that they, too, induce more deleterious effects in early development than in adulthood. For example, abnormalities in the reproductive system are induced by EDCs when animals are exposed embryonically or as neonates, but not when they are exposed as adults . Epidemiological studies have also shown that children exposed to EDCs early in life, even for a highly restricted period of time, may suffer significant and long-term consequences that arise later in life . Thus caution must be taken in assessing if particular EDCs (or EDCs at particular levels) are harmful if data is taken from adult populations (whether human or animal). Exposure to these compounds may be relatively benign in adults. However, even if present only transiently during a critical period of development, EDCs may induce significant detrimental effects that may not emerge until later in life.

EXPERIMENTAL MODELS FOR TESTING EDC EFFECTS

Given the expansive number of chemicals that need to be screened for EDC activity, what is the best experimental system to use? Several laboratories have utilised rapid screens in cell lines or in yeast to assess estrogen or androgen binding activity. These tests are fast, but they do not adequately address how EDCs will alter development of complex tissues, such as those comprising the central nervous system.

Numerous studies designed to investigate the effects of EDCs on development of peripheral reproductive structures have been carried out in rodents, and mammals provide an excellent system in which to assess how EDC exposure may interfere with human development. However, rodents and other mammals have limitations as an experimental system. Specifically, the number of pups obtained with each mating is small, the EDCs have to be administered by methods that do not mirror how organisms are exposed under natural conditions, and rodent development is relatively slow. Therefore, using rats and mice to screen the thousands of chemicals that are on the list of potential EDCs would necessitate a very long period of study.

Finally, while rodents provide arguably the best system in which to model EDC effects in humans, they may not provide the best system in which to address EDC effects on wildlife populations, in particular those that are aquatic. For example, it has been estimated that greater than 60 per cent of the alkylphenol polyethoxylates that are ingredients of nonionic detergents, paints, herbicides, and pesticides (which are produced at a rate greater than 300,000 tons per year) end up in aquatic environments where they may accumulate in both sediment and biological material. Moreover, environmental studies indicate that aquatic species are particularly sensitive indicators of the deleterious effects of EDCs.

An excellent model system, which provides both the ability to rapidly screen a large number of chemicals and to assess the effects of EDCs on complex vertebrate development, is the African clawed frog, Xenopus laevis ("Xenopus"). There are many advantages to using Xenopus as the experimental model. First, these frogs are totally aquatic, so the EDCs under study can be added directly to the water that the frogs live in, a situation that simulates how many wildlife populations are exposed to EDCs in the environment. Second, mating on a daily basis can be induced year round, thus a large number of embryos can be obtained: on the order of 100-1000 with each mating. Third, early development in these frogs is rapid with respect to other vertebrates: animals develop from a single-celled fertilized egg to a freely swimming tadpole in only 2.5 days. Finally, Xenopus is arguably the best understood preparation for studying molecular mechanisms underlying vertebrate development and is particularly amenable to studies of neurogenesis and neuronal differentiation.

THE BIOLOGY OF THE DEVELOPING NERVOUS SYSTEM IN XENOPUS LAEVIS

The developmental events that underlie formation of the nervous system are extraordinarily well characterised in Xenopus. Neurogenesis begins approximately 13 hours after fertilisation . As development proceeds, more new neurons are born, and they undergo a complex and highly regulated set of developmental changes that include: migration to appropriate places within the nascent nervous system; elongation of the processes called axons and dendrites that transmit and receive the electrical signals that are the coinage of information transfer in the nervous system; formation of chemical contacts called synapses between individual nerve cells and; expression of selective sets of neural-specific genes that allow specific subclasses to perform their appropriate functions (e.g., sensory neurons that receive information from the environment versus motoneurons that control muscle cells and movement). In particular, the generation and differentiation of primary sensory neurons that innervate the skin primary motoneurons that provide efferent control of axial (trunk) musculature, the formation of neuromuscular synapses, and the relationship of neuromuscular development to swimming behaviour are all highly stereotypic developmental programmes that have been thoroughly characterised at the level of the whole embryo. Moreover, the concomitant cellular and molecular changes that occur within single identified populations of neurons and muscle cells (myocytes) which underlie these developmental processes are just as reproducible and well-documented. This extensive understanding of normal development in Xenopus is of great advantage when trying to determine precisely which developmental processes are altered, deterred or aborted when animals are exposed to EDCs.

In addition to the wealth of literature describing development of the nervous system of intact Xenopus embryos, numerous studies have now shown that cells destined to become neurons or myocytes (but ones that have not yet adopted the defining characteristics of these specialised cells), can be isolated from the developing embryo and maintained in a dish as a dissociated cell culture (in vitro). Under these conditions, these cells will not only survive, but will go on to differentiate as neurons and faithfully reproduce many aspects of normal neural development, including elongation of processes, appropriate expression of ion channels that generate both electrical signals and transduce chemical signals at synapses, and the formation and maturation of synaptic contacts with appropriate targets (e.g., motoneurons will form synapses with muscle cells in vitro).

Thus, the Xenopus embryo provides the advantage of being able to observe effects of putative EDCs not only in the intact embryo, but also under in vitro conditions where the environment can be directly manipulated and controlled, and where the molecular actions of specific factors can be determined. For example, the conservation of developmental programmes extends to understanding how specific trophic signals (compounds released by developing cells including other neurons, as well as nonneuronal target cells) promote neuronal survival, neuronal differentiation, guide axon outgrowth and govern synaptogenesis. Because steroid hormones are known to regulate the expression of neurotrophic factors and neurotrophin receptors, interference with neurotrophin signaling pathways may be a likely mechanism by which EDCs could disrupt neuronal differentiation and synaptogenesis. In particular, preliminary data from our laboratory indicates that early exposure (prior to formation of the nervous system) to both the endogenous estrogen 17þ-estradiol, and to the EDCs, methoxychlor and nonylphenol, induces significant deficits in neural development, with the most notable changes observed in cells derived from part of the developing nervous system termed the neural crest. These neural crest cells require specific trophic factors for both survival and differentiation and consistent with this requirement, we have also shown that the ability of these trophic factors to induce differentiation of neurons developing in vitro is inhibited by these EDCs.

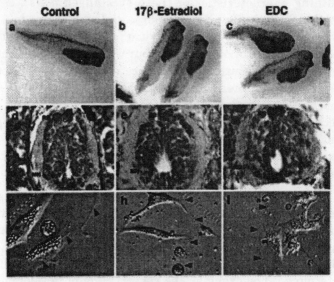

Fig. 19.1

CONCLUSION

These data suggest that Xenopus embryos can be used to rapidly and reliably screen for detrimental effects on vertebrate neural development, and that the ability to study neuronal differentiation both in whole embryos and in dissociated cell cultures makes Xenopus an excellent model system not only for screening potential EDCs for estrogenic and anti-androgenic activity, but for delineating the molecular mechanism of EDC action. With advances in this field, it is hoped that the dangers posed by EDCs to wildlife and to human populations will be fully realized so that further action can be taken to decrease environmental contamination.

Photographs in panels a-c show representative examples of tadpoles exposed to normal saline environment alone (control) or those exposed to the naturally occurring estrogen, 17þestradiol, or the EDC, nonylphenol. Animals were exposed at ~10 hrs after fertilisation (a time before nervous system tissue begins to form) and maintained in the hormone treatment for ~2 days. Both 17β-estradiol and nonylphenol had significant deleterious effects on the development of these tadpoles. Panels d-f show representative cross-sections through the spinal cords of control animals and those exposed to steroids or EDCs. Large sensory neurons (RB) and motoneurons (MN) can be identified in all animals, suggesting that the gross development of the central nervous system is not disrupted, but some of the motoneurons in the 17β-estradiol- and the EDC-treated animals seem pale and not healthy. Panels g-i show representative examples of muscle cells (m) and neurons (n and arrowheads) obtained from dissociating the part of the embryo that contains the developing spinal cord and some of the tail musculature at an early stage. These dissociated cell cultures provide a convenient system in which to test directly the effects of EDCs on the ability of specific cell types to survive and differentiate.

20
Applications of Mechatronics to Animal Production

INTRODUCTION

The concept of biosensors technology applied to animal production, mainly based on the miniaturised electronic mechanics (MEM) has being used since the mid 70s into several stages of production, such as feeding, detection of metabolic testing in animal husbandry, as well as to individual identification and monitoring, which is an important step towards tracking of actions and application of traceability of events and processes in the animal protein production chain. Last generation of such devices includes the real possibility of storing animal data as well as providing authentication protocols. The concept of specific management of a certain event rather then treating the herd/flock as a whole, likewise the precision farming, leads the precision animal production to re-evaluate losses and misdiagnosis by increasing the efficiency and accuracy and the use of precision techniques.

The application of mechatronics in animal production is found through the use of biosensors and MEMs, improving data collection and allowing more precise decision making actions.

This paper presents some examples of the use of this technology in specific areas related to animal production.

The overall performance of animal production depends on the herd or flock management as well as the nutrition, sanity control and lodging facilities. The concept of this kind of production is directly related to the reduction of selected losses and process control. Each production segment is controlled for reaching optimisation in the whole production system.

The concepts of precision animal production may apply at farm level for the animal management, housing environmental control, disease and nutrition control, information and identification, and ultimately, the overall treaceability In the last decade new technical tools have been introduced in animal production units/farms as support to decision making, especially for management, feeding strategies, animal health and fertility. Along with that, specific computerized systems were developed in order to elaborate the related variables and to provide the manager/farmer with opportune and appropriate tools and alert signals.

Average based models were the basis for the standard method in commercial animal production farms for monitoring most operation, and the forecasted values were compared to measured ones as well as to the next forecasted value, generating a predict average estimated value and, generally introducing an error, so called deviation. This has been applied for feeding system, reproduction practices as well as for predicting behavioural patterns.

With the advancement in microelectronics the possibilities of its use in animal production became feasible mainly for providing reduction of losses by better supporting decision making processes. Biosensor technology has great potential for improving animal welfare, health and production efficiency. The recent increasing incidence of diseases, such as bovine spongiform encephalopathy (BSE), tuberculosis, brucellosis, mastitis, and foot and mouth, has raised concern in the livestock industry and in society. Infectious diseases of livestock have major implications not only in animal welfare and production efficiency but also in human health, and food safety and quality.

According to Holroyd, the future of animal protein commerce depends mainly on an industry reacting towards the following concepts: honesty, openness, detailed information available, traceability, assurance of quality, and flexibility for changes. For the retailer or fast food buyer, it is only possible to build up a business when quality is always renewed, when final design is correct, and it is always available in the right place at the right time. Electronic traceability will enhance efficiency and accuracy in warranting safety in the animal's products chain.

This paper reviews the state of art as well as the prospective possibilities of the use of mechatronics in animal protein production systems.

Overall view on the possibilities of precision animal production Currently, diseases are controlled mainly with vaccines and drugs but the emergence of antibiotic-and drug-resistant pathogens means that diseases will continue to be a problem.

Moreover, the use of antibiotics will become even more severely restricted in the future.

Biosensor approaches are a promising tool to diagnose, and thereby aid in controlling animal disease in a more individual basis. These systems are inexpensive and reliable diagnostic tools that can be used by non-specialists. In addition, the test could be done at the animal's side providing immediate information about the status of a disease.

CHEMOTHERAPEUTICS IN ANIMAL HUSBANDRY

The widespread use of antibiotics and chemotherapeutics in animal husbandry (to control diseases and improve animal performance) has led to the occurrence of veterinary drug residues in foods of animal origin. Many countries have been introducing more restrictive food control measures but traditional microbial methods are not sensitive enough to meet new regulations and classical analytical techniques are often precluded owing to the level of experience, skills and cost required. Biosensor technology offers an alternative drug-screening method that is highly sensitive, does not require sample preparation, and can be rapidly carried out on-line at a low cost. Biosensors could also be used in the detection of metabolic levels in veterinary testing and animal husbandry, for example estrus detection by monitoring progesterone levels in milk By the mid-1970s experiments had been carried out with electronic transponders for individual feeding of cows and automatic data recording.

The electronic "black boxes" (first generation) were attached to collars used around the neck.

Later on, further miniaturisation of electronics allowed the development of tiny electronic transponders, which could be injected under the skin (second generation). Also the price declined dramatically. Because of the many logistical and tactical benefits of electronic animal identification, a worldwide market could arise for this application, primarily for agricultural animals, but also for companion and zoo animals. This needed the standardisation of codes and interrogation techniques. For this purpose, IOS came up in 1996 with two standards: ISO 11874 for the 64-bit code structure and ISO/11785 for the combined FDX/HDX interrogation protocol, working at 134.2 kHz. The third generation, currently under development, includes also read/write possibilities for storage of the (medical/genetics) history of the animal and sensor technologies for automatic monitoring of animal health and performance. Moreover, advanced third generation transponders can also be provided with authentication protocols to prevent fraudulent copying of transponder codes. IOS is also developing a standard for this new generation, which will be compatible with the existing standards.

It is understood as precision farming the use of special techniques and tools that permits specific management for a certain specific site and/or specific situation on field occurrence.

The use of such techniques and/or tools is supposed to lead management to a certain specific decision, and more precise action, rather than the use of average based values decision. Likewise in animal production large herds or flocks incorporate losses due to management decision based on average values. As an example, literature states that the upper critical housing temperature (UCT) for raising poultry breeders is around 28°C, however when studying individual animals it is found that UCT can vary from 27 to 32°C in the same genetics, and in some individuals the fluctuation within this range can cause losses up to 4 in the final result of fertile eggs. Apparently geneticists could detect through molecular genetics and genome tracing, a genetic marker responsible for this fluctuation in environmental response, leading the genetic development of breeders in a direction of a specific selection of thermal resistant animals and consequently, making the flocks more homogeneous, reducing losses related to heat stress exposure.

The use of electronic olfactometry is another example. Electronic olfactometry mimics the human smelling system by using an array of gas sensors and a simulated "brain", transforming the organic compounds contained in the headspace of the sample into an electrical signal in the form of curves via chemical sensors (metal oxides). Currently, the combined use of an array of gas sensors and neural networks approach provides a rapid method for measuring smells and a complement to the human nose in smelling analysis. In this sense, quality control in the food industry is one application that has seen the strongest development in recent years.

Gonzalez-Martin et al studied the differentiation of products derived from Iberian breed of swine by the use of olfactometry for classifying products on the basis of the diet they have received using adipose tissue as samples. The occurrence of fraudulent practices, consisting of the use of greased feed in an attempt to imitate the characteristics of range-diet animals, means that new methods are required to characterise these animals on the basis of their diet. Characterisation of the aroma of Iberian breed products has mainly focused on an algorithm that uses the electronic nose for identifying and classifying the specific meat product with precision. Olfactometry can be used as well to identify hazardous gases inside animal housing.

BIOSENSORS DEVELOPMENTS

Most biosensor developments have been in the biomedical field, where many in vivo applications demand small sizes. Claycomb and Delviche found that the on-line nature of a milk progesterone sensor did not require that the sensor be miniaturised to the point of utilising microfabrication technology. Their primary considerations were the physical sensor design, fluid transport, optical sensor configuration, fluid mixing, sampling of the raw milk, and automation. The ideal biosensor would be a probe, similar to that found in a pH meter.

Since the standard enzyme immunoassay identification of progesterone (EIA) is currently performed and standardised using microtitre plates, they used this technique as a starting point. The biosensor was then developed using EIA for molecular recognition and consequence identification of estrus in milking cows.

Radio frequency identification (RFID) plays a key role in electronic monitoring systems, which are inherently related to sensing systems. This combination makes it easier to switch from intensive to semi extensive animal husbandry systems (e.g. group housing of sows) as cited in Geers et al. Different systems are used as RFID.

Integration of on-animal sensing devices opens possibilities for the automation of sophisticated tasks such as health monitoring and reproduction status (estrus, pregnancy). Some examples are transponders equipped with a temperature sensor, as presented by Nelson, or in combination with activity tracking. The accuracy of these implanted temperature sensors is about $0.2°C$. As in most cases not the absolute values, but just the relative changes contain the significant information, the resolution must also be specified, and because of the digital representation of the temperature measurement, it is necessary to provide the result with at least one decimal place.

Sensor-based transducers also have been developed for monitoring the body temperature, the electrical cardiogram (ECG)signal and the pH value. These sensors have been used to monitor stress on piglets during transportation. The sensors are both the strength and the weakness of the monitoring concept. Typical performance aspects are the selectivity, the accuracy/resolution and long-term stability. In particular sensors with selective biointerfaces can cause stability problems. An example of this class of biosensors is the interface with the enzyme glucose oxidase (GOD) for glucose detection. Improvements have been achieved with immobilising techniques.

Because the sensor circuitry of these advanced devices requires a more or less continuous energy supply, small (mostly Lithium based) batteries have to be integrated in the transponder. Despite the application of very low power electronics, the lifetime of such devices is limited. One improvement to extend this lifetime is dual powering by using an internal battery for measurement and storage of data together with external an external radiating power source for transmission of data to the reader. Another alternative is the use of an external radiating powering source for both the interrogation and the semi-permanent activation of the sensor circuitry, thus enabling unlimited use of the sensor/transponder.

Accurate behavioural data for supporting optimal housing design The optimal design of animal housing requires meeting a large array of variables, specially when the welfare standards are faced. For instance ideal dry bulb temperature, which associated to relative humidity and black globe temperatures away from the thermoneutral zone may lead to undesirable environment and consequent losses in production.

The thermoneutral dry bulb temperature for breeders during the production stage lies between 22ºC and 28ºC. When the upper critical temperature (UCT) is reached, the latent heat lost by evaporation is highly affected by environmental relative humidity level.

The bird's UCT is influenced by the ventilation rate, the presence of cooling devices, and the temperature of drinking water. The bird's thermal regulation process in response to heat stressing conditions uses extra energy, leading to loss in productivity. Broilers in the first three weeks are more sensible to sudden weather changes requiring a more isolated building.

Optimum poultry production requires a housing environment that can offer well-distributed ventilation within especially for the last week bird's requirements.

The season's characterisation is related to solar declination, and the solar orientation of a building is then affected by the solar radiation flux intensity that reaches all the housing sides throughout the day. During the Winter at latitude of 40°S, for instance the North side of a building receives in average three times more solar radiation than the East and West sides.

While during Summer the North and South receive together only half of the solar radiation that reaches the Eastern and Western sides. In lower latitudes, as the case of São Paulo State, by the Tropic of Capricorn those differences are more enhanced and, during some clear sky Winters the Northern side is highly affected in term of incident solar radiation.

The location of a poultry housing regarding solar orientation is then of importance. Depending on the time of the year some side of the building will have more incident direct solar radiation as well as the diffuse radiation according to the sun movement. This will influence directly the total heat load inside the building.

Using RFID Nääs et al recorded poultry breeder's behaviour and related their behaviour to the environment characteristics using telemetry in small-scale model housing in two different solar orientations. Six female breeders were used in the models and they all had a transponder implanted. Four readers were used to record their movement within the smallscale housing placed on: nest, drinker, feeder and wall. The environmental parameters measured were: dry, wet bulb and black globe temperatures. Data were compared and the breeder's behavioural pattern according to environmental characteristics was determined.

Knowing in a more accurate way the behaviour of animals by using RFID, it will be possible to design better housing for intensive animal production.

Use of transponders for ID in animal production and the use of treaceability processes.

Nevertheless consumers are quite aware of the health problems the ingestion of unsafe foodmay bring to them and their families and associate this item to the animals housing and management, ingestion of drugs and ultimately the process and conservation of the product throughout the market chain. Safety is one of the most demanded qualities in food products nowadays, and it interacts mainly for assuring quality. It is important to meet the consumer's requirements for food safety through the use of treaceability of the animal welfare and health control as well as the labor welfare and health, reducing also the risk of contamination from all sources in the process of production.

Devices for electronic animal identification, becoming available in the mid-1970s, facilitated the implementation of sophisticated livestock management schemes. The standardisation by IOS of the next generation of injectable electronic transponders opened a worldwide market for all species of animals. The third generation, currently under development, includes also read/write possibilities and sensor technologies for automatic monitoring of animal health and performance. The addition of these sensors facilitates the automation of sophisticated tasks such as health and reproduction status monitoring The discussion about the best place for implanting transponders in some species of animals is still being updated. Pandorfi et al and Silva et al presented solution for transponders applied in new born piglets inside the ear's base, for use in posterior traceability inside the whole swine production system. The use of transponders for complete electronic traceability in swine production remains a challenge. The use of electronic ear tags is still the most common way of tracking swine within the production sites.

Electronic identification of cattle usually referred to as RFID has many advantages for farm management. First, it can be regarded as a considerable improvement in relation to visual identification of numbers. The main advantages are the elimination of labor costs and the decrease of incorrect readings from 6 per cent to 0.1 per cent (Artman, 1999). RFID also facilitates the use of automated housing systems and combines the advantages of the conventional loose housing systems (relative freedom for the animals, attending some animal welfare demands) with the advantages of the stanchion barns (control of single animals). Allowing the automation of, for example, feed monitoring and rationing, weighing and drafting can implement sophisticated livestock management schemes.

Using RFID cattle management can be carried out on basis of the individual animal performance recording, dispensing of feed, geographic routing dependent on the animal status. Examples are robot milking and the implementation of geographic information systems to assess the potential transmission of infectious diseases between herds. Also from the point of view of return on investment, RFID systems seem to be a good solution. Other important applications enabled by injected electronic transponders are improvement of disease control and eradication as well as fraud control. The latter application is very important within the European Union (EU), where premiums are being paid to stimulate extensive sheep and beef production. Also within the EU, it is not longer allowed to eradicate some contagious diseases by means of vaccination. In case of an outbreak, it is very important to trace back the origin, movements and contacts between animals to be able to stop the further dissemination of contagious diseases.

In practice, RFID implementations can give rise to several problems. Reading speed and distance must be optimised for specific applications. The International Committee for Animal Recording (ICAR) developed in 1995 a set of requirements regarding (among others) the reading distance and reading speed. Other issues include biocompatibility of encapsulation, as well as the injection site in connection with migration problem, recovery in slaughterhouses, and the open trade that needs standardisation, and proper effective management of issued unique life-numbers.

Due to the consumers demand on requiring certain characteristics of the product including those related to sanity control up to ecological features, identifying the final product is one of the objectives on the treaceability process. The treaceability in this case is to assure the consumer herd or flock welfare and health, good nutrition, non use of certain drugs and that there is adopted in the farm environmentally safe waste management.

DETECTION OF ESTRUS IN DAIRY COWS

In many countries breeding results are declining as can be illustrated by a decrease in conception rates after artificial insemination and by an increase in calving intervals. Part of these problems seems to be related to the failure to detect estrus or to the misdiagnosis. It is generally accepted that heat detection efficiency is lower than 50 per cent in most dairy herds (and, considering progesterone concentration in milk or plasma on the day of service, from 5 to 30 per cent of the cows were not in or near estrus when inseminated.

Continuous or almost continuous observations studies have shown that displays of estrous behavior occur unevenly throughout each 24 h period and many of them are of short duration. These findings were recently confirmed on large number of animals using electronic devices that allow continuous monitoring of behavioural activity. Therefore regular periods of observation of estrus are required which is less and less possible within the constraints of actual management practice, particularly because individual dairy herds increasing in size, the manpower input per cow will decrease

Tested by Saumande, the DEC system (IMV Technologies, France) is an electronic device designed to detect estrus in the bovine species. Its principle is based on the electronic detection of standing mounts accepted by cows in estrus. The criteria (number of, length and interval between mounts) are analysed by a microprocessor associated with the sensor and give a definition of the onset of estrus. During the course of this experiment, the efficiency of estrus detection by visual observation (69.8%) was higher than the efficiency usually reported in the literature (38 to 56%); however these values are averages and efficiencies over 60 per cent have been already observed in individual herds. The continuous (24 h a day) surveillance of the cows represents one of the main advantages of electronic devices when detecting estrus.

Therefore, it looks surprising that the efficiency of the DEC system was approximately only 50 per cent of the efficiency provided by visual observation. Their data were not in agreement with this statement as the false positives represented 12.8 per cent of the estrus detected by the DEC system versus only 7 per cent of those detected by visual observation. In comparison with the accuracy reported for HeatWatch, another electronic pressure-sensing system, the accuracy for the DEC system (84.6%) seems to be lower than the value of 100 per cent reported by Xu et al but similar to the 85 per cent observed Stevenson et al.

Another way of detecting estrus in milking cows is the use of on-line system to automatically monitor luteal function by assay of each cow's milk for progesterone every time the milking machine was attached. Claycomb and Delviche present a sensor technology to implement the rapid assay and create an automated sensing system for operation in the dairy parlor during milking, and to evaluate the sensor performance and reusability for multiple test cycles. For an automated on-line sensing system, sampling of the raw milk stream was necessary. Although various sampling devices are commercially available to remove small volumes from a flow stream, a device was specifically designed for this biosensor. Since this mechanism is in contact with the milk being shipped commercially, cleanliness is of considerable importance. By using pinch valves, the milk never wetted the valves, and sample flushing by the next cow's milk was relatively quick and simple. Since flow through this valve was always inward, milk contamination was not a problem. The accumulated blocking effect of the milk proteins limited conjugate absorption to the fluidics components. The sensor was successfully developed to work on-line in a dairy parlor using a control computer for sequencing its operation.

Reduction of losses in animal production by the use of mechanised and electronic processes

Individual electronic feeding has been used in dairy cows and grouped gestating sows breeders for nearly ten years. A transponder is placed in a necklace that opens automatically the gate and feeds individually each animal according to its milk production and need. The use of this technology has been developed mainly for large mammals (beef cattle, dairy cows and swine breeders) and is yet in development for broilers, specially breeders. Processes such as milking of cows in the milking parlor are dependent on the control of the pressure in the pump, and the individual monitoring is important for identifying variables that may affect production.

In Nääs and Fialho is presented a system that uses a radio transmitter and a flux register inside an on-line milk production, pressure data can be measured and controlled, reducing the incidence of mastitis (due to pressure fluctuation) in the herd.

CONCLUSION

Biosensors show great potential for applications in the livestock industry, particularly where rapid, low cost, high sensitivity and specificity measurement in field situations is required, but the technology must overcome several obstacles before becoming a commercial success. The barriers for the slow transfer of technology from the lab to the field are mainly technical, particularly in the production methods to fabricate reliable and inexpensive sensors, in the stabilisation and storage of biosensors and above all in the total integration of biosensor systems. The application of biosensors requires a specification that should include a sampling system, a biosensor, a calibration system and a model of how the information is to be used to control the process of interest.

Applications of Mechatronics to Animal Production

More than 100 million dairy cattle, as well as sows, in the developed world are managed using artificial insemination and thus the potential for the progesterone biosensor system is tremendous even if only a small proportion of farmers take up this technology. The biosensor system will also be capable of being extended to disease monitoring, analysing routinely for markers of mastitis infections or antigens of disease.

With the use of miniaturised electronic mechanism (MEM) will be possible to record and control, each time in a more accurate way, events or diseases in order to respond for the optimisation of animal protein production.

The use of automation/mechatronic in animal production will help farmers decrease losses during the animal production cycle, by the use of precision principles and more accuracy, improving animal management. The role of traceability in the animal protein production process remains a challenge for facing the consumer's demand, while practical solutions in the complete food chain are still missing. There is a large room for transfer of technology as well the development of new devices and applications of new techniques and systems.

Index

A

Anaerobic digestion in animals, 127-133
    application, 131-132
    digestor design, 129-131
    introduction, 127-128
    odour impact, 128-129

Animal cell biotechnology, 141-147
    exploration on animal model for senile memory deficits, 143-144
    gene transfer in animals, 144-147
    immortalization of cells in culture, 142-143
    introduction, 141

Animal cloning, 64-75
    can organs be cloned for use in transplants, 73-74
    development of animal cloning in the lab, 65-66
    introduction, 64-65
    process of animal cloning, 66
    types of cloning technologies, 67-72
    what animals have been cloned, 72
    – are the risks of cloning, 74-75

Animal genetic resources for agriculture and food production, 148-161
    considerable genetic variation among breeds, 158-159
    diversified use of livestock, 157-158
    diversity in animal genetic resources invaluable for future developments, 158
    food security and livestock-keys to poverty alleviation, 149-155
    introduction, 148-149
    livestock revolution underway, 155-157
    new approaches needed for sustainable livestock improvement, 160-161
    why worry about loss in genetic diversity, 159-160

Animal monoclonal antibody, 162-177
    monoclonal antibodies for cancer treatment, 170-171
    alemtuzumab, 174-175
    applications, 169
    gemtuzumab ozogamicin, 174
    hybridoma cell production, 167-169
    introduction, 162-163
    monoclonal antibody technology, 163-165

palivizumab, 175-176
production of monoclonal antibody, 165-167
recombinant, 169
types of monoclonal antibodies, 171-172
- Abciximab, 172-173
- Alemtuzumab, 174
- Basiliximab, 172
- Daclizuman, 173
- Gemtuzumab, 174
- Infliximab, 171
- Palivizumab, 175
- Rituximab, 175
- Trastuzumab, 176-177

Animal proteomics, 134-140
- antigen presentation studies using mass spectrometry techniques, 137
- application of proteomics to studies on phylogeny and evolution, 136-137
- applications on proteomics to animal physiology, 140
- cell proliferation, 139-140
- genomics and proteomic, 139
- introduction, 134-135
- protein database of drosophila, 135
- proteomics on the totipotent planarian stem cell, 137-138

Animal Welfare Assurance, 163

Applications of mechatronics to animal production, 188-201
- biosensors developments, 193-198
- chemotherapeutics in animal husbandry, 190-192
- detection of estrus in dairy cows, 198-200
- introduction, 188-190

Arber, Werner, 120

B

Biotechnology for animal breeding, 85-93
- future, 93
- genetic variation, 88-89
- introduction, 85-87
- molecular genetics, 87-88
- – – and animal breeding, 87
- new scientific technologies, 92-93
- selection for a strong immune response, 89-91

BOS Taurus genome, 113-114

Breeding of transgenic animals, 46-51
- how are transgenic animals produced, 47-49
- how do transgenic animals contribute to human welfare, 49
- industrial applications, 51
- introduction, 46
- medical applications, 50-51
- reasons for transgenic animal production, 47
- what is a transgenic animal, 46-47

C

Cloning with somatic cell nuclear transfer, 15-38
- artificial insemination, 18
- introduction, 15-16
- IVP embryos, 18
- materials and methods, 16-17

Index

morphometric analysis, 21-38
NT embryos, 17-18
pregnancy rates, 20
recipients of IVP or NT embroys, 18
Cloning, 64-75

D

Digestor design, 129-131

E

E. coli, 126

F

FAO State of the World Animal Genetic Resources Report, 102
Farm animal diversity, 99-105
hotspot of breed diversity loss, 101-102
introduction, 99-100
maintaining genetic diversity of livestock, 102-105
management of animal genetic resources, 100
protecting our common heritage, 102
some breeds are more equal, 100-101
Farm animal genomics, 106-119
BOS Taurus genome, 113-114
DNA and protein sequences data banks, 107-109
ethical reservations of farm animal genomic study, 118-119
farm animals, 109-113
fish genome project, 115-116
forgotten rabbit, 116
genome policies for transgenic animals, 118
introduction, 106-107

potential of farm animal genomics, 119
quantitative trait loci and genetic linkage, 117-118
sheep genome project, 115
swine sequence, 113
Food and Agriculture Organisations, 153, 155

G

Gene knockout technology, 8-14
fish embryo cell cultures for targeted gene inactivation, 12-14
objectives, 8-10
parameters affecting the efficiency of targeted mutagenesis in bovine cells, 8
targeted mutagenesis, 10-12
Genetic engineering, 76-84
ethical issues, 83-84
genetic modification of farm animals, 79-80
introduction, 76-79
production of modified food-producing animals, 80-82
Genetic resources for agriculture and food production, 148-161

I

In vitro fertilisation and cell culture, 39-45
advantages, 44
cell culture, 42-43
features, 44
growth cell culture, 43-44
introduction, 39-40
IVF cycle, 40-42
organ culture, 44

techniques of organ culture, 44-45
International Food Policy research Institute (IFPRI), 155
International Livestock Research Institute, 155
Introduction, 1-7
    application of animal biotechnology, 1-4
    gene knockout technology, 4-5
    somatic cell nuclear transfer, 5-7
    transgenics, 4

M

Morphological abnormalities in animals, 178-187
    biology of the developing nervous system in Xenopus laevis, 184-186
    cellular and molecular mechanism underlying EDC effects, 180181
    critical periods in development, 181-182
    endocrine disrupting compounds, 178-180
    experimental models for testing EDC effects, 182-184
    introduction, 178
Muller, Alexander, 99

N

Nathans, Daniel, 120
National Agricultural Research Systems, 103
National Cancer Institute, 62
National Institutes of Health, 162
New Zealand Animal Welfare Act of 1999, 16

P

Priorities in animal biotechnology, 94-98
    animal models for the study of human health, 94-95
    development of rapid diagnostic kits, 96-97
    introduction, 94
    livestock industries, 95-96
    priority areas identified for animal biotechnology, 97-98
Production of Infertile Aquatic Species, 6
Public Health Service, 162

R

Recombinant DNA technique, 120-126
    chemeric plasmids, 123-124
    cloning and relation to plasmids, 121-123
    introduction, 120-121
    synthetic insulin production using recombinant DNA, 124-126

S

Smith, Hamilton, 120
Stevenson, 199

T

Transgenesis and gene therapy, 52-63
    different types of viruses used as gene therapy vectors, 59-61
    embryonic stem cell-mediated transgenesis, 54-56
    gene therapy, 53-54

introduction, 52-53
recent developments in gene therapy research, 61-63
transgenic astrocytoma models, 58-59
– models of astrocytomas, 56-58

U

UCL Institute of Ophthalmology and Moorfields Eye Hospital NIHR Biomedical Research Centre, 61

United Kingdom, 173

United National, 155

University of California, 63

USDA Biotechnology Risk Assessment Grants, 7

W

Whitehead Institute for Biomedical Research in Cambridge, 74

World Animal Genetic Resources for Food and Agriculture, 99

Z

Zenapax, 173